Understanding Practice in Design and Technology

DEVELOPING SCIENCE AND TECHNOLOGY EDUCATION

Series Editor: Brian Woolnough,
Department of Educational Studies, University of Oxford

Current titles:

John Eggleston: *Teaching Design and Technology (Second edition)*
Richard Gott and Sandra Duggan: *Investigative Work in the Science Curriculum*
David Layton: *Technology's Challenge to Science Education*
Michael Poole: *Beliefs and Values in Science Education*
Keith Postlethwaite: *Differentiated Science Teaching*
Michael J. Reiss: *Science Education for a Pluralist Society*
Jon Scaife and Jerry Wellington: *Information Technology in Science and Technology Education*
Joan Solomon: *Teaching Science, Technology and Society*
Clive Sutton: *Words, Science and Learning*
Brian Woolnough: *Effective Science Teaching*

Understanding Practice in Design and Technology

RICHARD KIMBELL
KAY STABLES
RICHARD GREEN

Open University Press
Buckingham · Philadelphia

Open University Press
Celtic Court
22 Ballmoor
Buckingham
MK18 1XW

and

1900 Frost Road, Suite 101
Bristol, PA 19007, USA

First published 1996

A catalogue record of this book is available from the British Library

ISBN 0–335–19554–7 (pb)   0–335–19555–5 (hb)

Library of Congress Cataloging-in-Publication Data
Kimbell, Richard.
    Design and technology activities: understanding practice/
Richard Kimbell, Kay Stables, Richard Green.
      p.   cm. – (Developing science and technology education)
    Includes bibliographical references and index.
    ISBN 0–335–19555–5 (hbk)   ISBN 0–335–19554–7 (pbk)
    1. Technology – Study and teaching (Elementary) – Great Britain.
2. Design, Industrial – Study and teaching (Elementary) – Great Britain.
I. Stables, Kay, 1954–   .   II. Green, Richard, 1958–   .
III. Title.   IV. Series.
T107.K58   1996
372.3′58044′90941 – dc20   95–42150
                                                                  CIP

Typeset by Type Study, Scarborough
Printed in Great Britain by St Edmundsbury Press, Bury St Edmunds, Suffolk

Contents

Series editor's preface

It may seem surprising that after three decades of curriculum innovation, and with the increasing provision of centralised National Curriculum, that it is felt necessary to produce a series of books which encourage teachers and curriculum developers to continue to rethink how science and technology should be taught in schools. But teaching can never be merely the 'delivery' of someone else's 'given' curriculum. It is essentially a personal and professional business in which lively, thinking, enthusiastic teachers continue to analyse their own activities and mediate the curriculum framework to their students. If teachers ever cease to be critical of what they are doing, then their teaching, and their students' learning, will become sterile.

There are still important questions which need to be addressed, questions which remain fundamental but the answers to which may vary according to the social conditions and educational priorities at a particular time.

What is the justification for teaching science and technology in our schools? For educational or vocational reasons? Providing science and technology for all, for future educated citizens, or to provide adequately prepared and motivated students to fulfil the industrial needs of the country? Will the same type of curriculum satisfactorily meet both needs or do we need a differentiated curriculum? In the past it has too readily been assumed that one type of science will meet all needs.

What should be the nature of science and technology in schools? It will need to develop both the methods and the content of the subject, the way a scientist or engineer works and the appropriate knowledge and understanding, but what is the relationship between the two? How does the student's explicit knowledge relate to investigational skill, how important is the student's tacit knowledge? In the past the holistic nature of scientific activity and the importance of affective factors such as commitment and enjoyment have been seriously undervalued in relation to the student's success.

And, of particular concern to this series, what is the relationship between science and technology? In some countries the scientific nature of technology and the technological aspects of science make the subjects a natural continuum. In others the curriculum structures have separated the two, leaving the teachers to develop appropriate links. Underlying this series is the belief that science and technology have an important interdependence and thus many of the books will be appropriate to teachers of both science and technology.

Richard Kimbell has been one of the most perceptive, influential and significant figures in the development of school technology throughout the world. The researches that he and his colleagues have done, especially with the APU and the ESRC's Understanding Technological Approaches projects, have been seminal in defining technology, in analysing what students and teachers do, and in developing methods of evaluating and assessing their work. This book brings together the insights from such work and shows, in a readable and profoundly perceptive way, how design and technology can be learnt and appreciated. It also, by concentrating on the individual strengths and potential of each and every student, shows differentiation as a creative and liberating process rather than a problem. It is an exciting and stimulating book, all teachers of technology and of science will be able to gain fresh insights into their own teaching from it.

We hope that this book, and the series as a whole, will help many teachers to develop their science and technological education in ways that are both satisfying to themselves and stimulating to their students.

Brian E. Woolnough

CHAPTER 1

Defining the task

An uncertain tradition

This book is unusual in technology since it is based on research. This does not mean that it is abstract and convoluted but rather that its central messages have been derived from painstakingly detailed enquiries. The focus of these enquiries has been the classrooms, studios and workshops where Design and Technology is being taught. Specifically, the research that informs this book has sought to unpick and examine some of the interconnections between children's learning in Design and Technology and the practice of teachers. It is perhaps inevitable that these enquiries have raised some fundamental questions about the nature of learning in Design and Technology and about its unique role in the curriculum. Again, not surprisingly, this has led to some heart-searching debate about how we are to define Design and Technology: how we delineate the boundaries between what is (and what is not) a Design and Technology activity.

We are all too well aware that this debate has been going on at the macropolitical level simultaneously with our own efforts. No teacher can be unaware of the impact of National Curriculum policy, and specifically Design and Technology teachers (and all primary teachers) are painfully aware of the impact of changing National Curriculum policy for Design and Technology since 1990. Whilst we might retain little sympathy for politicians as a whole, they do – on this matter of Design and Technology – deserve a little sympathy, or at least a little tolerance, since at the outset of the great National Curriculum adventure (in 1988) there was painfully little foundation on which to build a coherent and progressive Design and Technology programme. At that time, one of the few things upon which writers in the field were agreed, was that there was a desperate lack of research to inform decision-making.

> . . . Design & Technology lacks a research base in pupils' understanding and learning such as is available in the cases of mathematics and science . . .
>
> (DES/WO, 1988)

> Craft Design & Technology stands out as the most under-researched area of the curriculum. The literature of the subject barely exists.
>
> (Penfold, 1988)

We also need to remember that 25 years before the introduction of the National Curriculum, in the early 1960s, the subject did not even exist in anything remotely like its present form. There was plenty of craftwork and bits of applied science, but nothing identifiable as Design and Technology as we currently know it. Twenty-five years is an astonishingly short time in which to evolve a whole new curriculum discipline, particularly when the vast majority of the curriculum has remained so stable for so long. As Williams somewhat acidly observes:

> The fact about our present curriculum is that it was essentially created by the 19th century, following some 18th century models and retaining

elements of the mediaeval curriculum near its centre.

(Williams, 1965)

Whilst this was a comment about the 1960s' curriculum, there are many observers who argue that the new National Curriculum is following precisely in these traditional footsteps. But everyone acknowledges that the exception to the rule is technology. They may not know why it's there – or what it uniquely can contribute to the education of our young people – but everyone can see that it is new. It is because of this newness that it is so unstable, and this instability makes it vulnerable. The many versions of National Curriculum technology that have done the rounds in the last five years or so are ample testimony to the uncertainties that surround it.

This book – and the research on which it is based – represents an attempt to shed some light on the issues that surround Design and Technology and its practice in schools.

Research sources

There are two principal research sources for the materials in this book. The first is the Assessment of Performance Unit project in Design and Technology 1985–91 (hereafter called the APU project) funded by the Department of Education and Science and based at the Technology Education Research Unit in Goldsmiths University of London. This was a project within which we explored problems and approaches to assessment in Design and Technology. The brief was to develop tests of Design and Technology capability and (in 1988–9) to administer them to the nation's 15-year-old cohort. The sample was to be 2 per cent of the whole cohort (i.e. approximately 10 000 pupils in 700 schools throughout England, Wales and Northern Ireland) and was to be largely a random sample. As this survey pre-dated the National Curriculum, a significant section of the sample were not taking any technology-related courses at the time of testing. By virtue of the size of the sample, it provided extremely reliable data that supported generalisable conclusions. Most of the testing was

based on short-term, pencil and paper test responses, reducing the ability of the research to comment on the interaction of the tests with practical capability, but a sub-sample of the survey was used to examine this issue. We ended up with a three-pronged assessment framework that linked short pencil and paper tests with longer, more practical and more 'real' activities and we designed a survey that allowed us to cross-relate performance between them. In the end, for reasons that will become apparent, the short tests provided some remarkable insights both into the capability of pupils and into the nature of the activity. None the less, these findings also provoked the need to examine in more detail the longer-term 'real-time' experience of Design and Technology in the classroom.

In the later years of our APU work, we became increasingly interested in the *development* of capability and this required us to examine it at a range of points in children's schooling. Accordingly, we sought funding for a project from the Economic and Social Research Council, and in 1992 we established a new project, Understanding Technological Approaches (hereafter called the UTA study) again based at the Technology Education Research Unit in Goldsmiths University of London. The rationale behind this project required us to explore *models of practice in technology* by developing 80 case studies of pupils working on technology projects in 20 schools in the Greater London area. In order to address issues about the development of capability, the case studies deliberately span all four key stages. The methodology was broadly observational with a data collection system based on trained observers watching (literally) every minute of the pupil projects and recording what happened in every five-minute session. The observational data are enriched with discursive data from interviews with the teachers and pupils, and with performance data as a result of assessing the work. The use of a common observation framework across the key stages enabled us to compile data and report on progression and continuity issues in ways that have never before been possible. However, with only 80 case studies over four key stages, the sample is small and does not claim to be representative. The findings are therefore not necessarily generalisable.

These two projects therefore provide a complementary background to this book. The APU project was very large scale and produced very reliable and detailed data – but based largely on limited measures (short tests) of capability. The UTA study used a small sample but examined the reality of the whole process of design and development as it operated for 80 pupils across four key stages. Taken together, they enable us to be confident that where the findings of the two projects are in alignment, we have discovered something of real significance. In order to maximise the possibility of making these connections, we developed the UTA observation framework from our earlier work on the APU assessment framework. It had proved very useful and robust as an assessment framework for the APU project and has since proved equally effective within the longer-term observations of the UTA study.

Defining the areas of confusion

The two projects have thrown up a significant number of problematic issues about which too little is known and on which (if our findings are to be believed) a great deal hangs. We shall briefly outline them here, raising an agenda that we shall seek to respond to through subsequent chapters of this book.

The context of tasks

Real tasks do not exist *in vacuo*. They exist in real houses or gardens or shops or car parks or hospitals, and the *setting* of the task is a major determinant of the *meaning* of that task. If you were invited to 'design a door handle' it would have very little meaning until you could see the context for which it is intended. It might be for a child's playhouse, or for an industrial kitchen, or a heavy goods vehicle. In each case the issues that the designer needs to consider are to a large degree defined by the *context*. Equally, the success of the

outcome can only be determined by examining its operation in the same context.

Serious criticism has not infrequently been levelled at formal schooling in that it so frequently sets learning into artificial situations, devoid of real meaning and creating an 'alien culture which lacks relevance to the everyday problem-solving practices and thinking which takes place outside it' (Hennessy and McCormick, 1993). Technology has the obvious opportunity to overcome this difficulty – since as designers and makers our field of operations is necessarily to do with the real world of products.

The idea that 'contexts' might be important for technology tasks was made formal and explicit in the APU project, where we created a series of eight-minute videos to capture the range of contexts within which our tasks were embedded:

- a pre-school playgroup
- a post office
- an elderly person's kitchen
- a backyard
- an industrial production line
- a design studio.

All eight videos are available on a single tape from Kent Education Television in Dover.

This 'hands-off' approach to context – through a video – is generally recognised as less powerful for children than experiencing it directly, which teachers often attempt to arrange for the projects they run.

However, what starts out as a transparently good idea can come seriously unstuck when enshrined in the legal verbiage of statutory curriculum legislation, and this is what happened to 'contexts' in the Design and Technology National Curriculum Order:

> We place much importance on pupils' design & technology activity being undertaken in a variety of contexts and specify what we have in mind by reference to home, school, community, business and industry, recreation . . .
>
> (DES/WO, 1990)

Not only were the contexts specified in a confused form (mixing up obvious location ones like 'home' with less obvious ones like 'recreation') but further it was implied that pupils should, of their own accord, be able to discern technological tasks from within these contexts. This caused serious confusion for many teachers as we shall see in Chapter 4. It damaged the credibility of teaching through contexts and subsequently led to their elimination from later National Curriculum documents.

This was all a great shame since the data from our two projects is quite clear on this point. The first critical message from our APU testing, and from observed UTA projects is that contextualised tasks provide richer learning experiences for children. This is for two reasons. First, because the context provides meaning for the task and second because it provides (in a very concrete manner) a series of trigger points for action – things that children can immediately get on and find out about or experiment with. Contextualising activities for pupils is a good thing. Contexts are enormously empowering for teachers and pupils alike.

However, they need to be used with professional care and expertise. And *how* to do this – and the strengths and weaknesses of doing it one way as opposed to another – are all quite proper subsequent questions, and we shall examine them later.

The hierarchy of tasks

Regardless of contextualising, the level at which the task is set has a serious impact upon its 'do-ability'. The APU data demonstrated that the subject matter of the task (e.g. electronic alarms or fabric constructions) counts for relatively little in determining how well pupils are able to perform. But it matters a great deal whether the task is set *loosely* or *tightly*.

In some tasks, we provided little more than a generalised suggestion that a given context was full of potential problems that could usefully be pursued by a designer; for example, 'think about issues of *protection* in this context . . .' We then left the pupils to identify for themselves the specifics of the tasks they might pursue. In other tests we provided a more comprehensive framework for the task by adding more precise demands to it. In the most

tightly specified tasks we went yet further and added quite specific design requirements to their task; for example, '. . . in the context of designing for backyards and small gardens; design a modular, self-watering, stackable, plant- holder system that will . . . etc'. Here was a quite specific starting point that provided little room either for negotiation or for confusion. We eventually evolved three levels of task that were hierarchically related, and the same structure can be applied to any context you care to mention.

- Contextual task: very open
- Framed task: some constraints
- Specific task: tightly defined

It is easy to understand the differences of demand that exist within these layers of task setting. The contextual level provides plenty of opportunity for pupils to 'take ownership' of the task, because they are – in a very real sense – deciding for themselves what the task will be. But there is very little support in it; it is not clear exactly what pupils should get on and do, and it might be easy to spend a long time unproductively trying to find a task with which one can make a start. At the other extreme is the reverse problem. It is quite obvious what the task is – and hence quite apparent how one might immediately proceed, but it is much more difficult for pupils to generate any personal ownership of the task. What do we do with the pupil whose reaction is '. . . I'm not interested in, I don't want, and I don't need a stacking plant holder . . .'? In this circumstance, the activity is hardly likely to generate much of a response from the pupil.

The big question, of course, is how these different layers of task-setting affect pupils' performance and again this is a matter to which we shall return in detail in Chapters 4 and 6.

Structuring the activity

Assuming that any given length of time is available for an activity, the use of that time can be managed in any number of ways. It might be planned only very loosely – with large blocks of time available to the pupils and only limited interventions by the teacher to steer the activity. Or it might be structured much more heavily – with more interventions and instructions by the teacher.

As with the hierarchy of tasks, the strengths and weaknesses of these two extreme approaches to structuring the activity are obvious in retrospect. In tightly structured activities there is plenty of support and few pupils lose their way in the activity. By the same token, however, there is little opportunity for pupils just to get on with it. In the more open structure it is much easier to waste time – or lose one's way completely – but those that stay on task can get on without interruption.

Pupil autonomy

The question that lies at the heart of this matter of the sub-structure of tasks is the extent to which pupils are able – and the extent to which they can be trusted – to get on by themselves. It is the issue of pupil autonomy. It is an issue about which we have been astonished as we examined the data emerging from the UTA study. At a common sense level one might expect that pupils start off dependent upon their teachers – but gradually they grow to become independent and autonomous learners. We would therefore expect to find evidence of this gradual transition of responsibility across the four key stages. But we did not.

The evidence is in fact quite different. There appear to be two quite distinct layers of transition. Through Key Stages 1 and 2 the pupils grow to become ever more independent so that at Year 6 the classroom is full of semi-autonomous learners using their teacher (and anyone else who happens to walk into the room) as a resource for their work. In Year 7 there is a total transformation with pupils reverting to a level of absolute dependency on the teacher. The style of operation in Year 7 classrooms and workshops must come as a real shock to these youngsters, and the clearest manifestation of it is the ubiquitous queue of children waiting to see the teacher. One almost never sees this in Years 5 and 6. Thereafter, throughout the last year of Key Stage 3 and through Key Stage 4 there is again a steady transition towards more autonomous activity, and interestingly there is a close resemblance between the teaching and learning styles of Key Stages 2

and 4. This is an issue to which we shall return in Chapters 5 and 7.

Closed or open tasks and tight or loose structures. These two generic features of Design and Technology activities both relate to the question of pupil autonomy. They have a strong tendency to interact with each other in ways that have a marked impact on the performance of different sub-groups of pupils. We shall explore different facets of these issues in Chapters 3, 4 and 6.

The iteration of action and reflection

In any Design and Technology activity, there is a careful balance to be struck between *active* designing (drawing, modelling, making, etc) and *reflective* appraisal of where we are going (identifying issues for judgement, identifying strengths and weaknesses in the work, etc). In the test development phase of the APU project it became obvious that the best levels of pupil performance were associated with activities in which action and reflection were kept in balance. Design and Technology is about the active pursuit of real problems, but it must be focused and directed by continuous awareness of the needs to be met, the priorities of the users, and the strengths and weaknesses of the work so far. In a Design and Technology task (as probably in any task) the relationship between action and reflection is iterative. Action forces issues into the daylight, and in reflecting on these issues, we raise further directions and possibilities for action.

To get a measure of the significance of this phenomenon, consider the difference between the order of events of the sub-tasks of these two 'evaluation' activities.

Activity (a)

Contextualising the use of the product
Identifying criteria of judgement for the product
Examining the product
Identifying strengths and weaknesses of the product
Redesigning to improve the product

Activity (b)

Contextualising the use of the product

Examining the product

Design development of a new product to be better than the original

Identifying criteria of judgement for the original product

Identifying strengths and weaknesses of the original product

Both structures have the same five elements, but they arise in a different order. The major difference is in the position of the *active* designing (Stage 5 or Stage 3), as opposed to the *reflective* identification and consideration of criteria. This has enormous consequences for pupil performance, especially in terms of the richness of the criteria subsequently identified as important. If pupils are able to think about these criteria having already tussled with the design challenge, then they are more aware of what is important and what is not important. The research data are quite clear on this point, and the interdependence of thinking and doing in Design and Technology enables us to understand why it is so. We shall explore this point more fully in Chapters 2 and 3.

Collaboration and discussion

Perhaps the most startling outcome from the development of the APU modelling tests was our realisation of the role and importance of group discussion. From the earliest trials we were aware of the impact that it had on helping pupils to crystallise their design ideas, and in the final form of the test activities we deliberately set a 20-minute discussion into the middle of the design and development phase. In turn, each of six pupils was asked to say (in two minutes) what they had done, why, and what they proposed to do next. It was then open for peer comment around the table – with the chairperson (the test administrator) being strictly constrained to neutral questions as prompts if necessary. It seldom was necessary, and in the vast majority of cases the interaction proved immensely significant in helping pupils forward in their designing.

One of our test administrators commented:

> This was a strategy that I had previously not put any emphasis on in my own teaching and I found it by far the most useful device for helping pupils extend their ideas. The pupils' response to each others criticism was a major force in shaping the success or failure of the artefact in their own eyes. Pupils saw this as a very rewarding activity and would frequently modify the direction of their own thinking as a result.
>
> (Kimbell *et al.,* 1991, section 9)

This was a common reaction among the administrators. Even though we had selected the administrators on the basis that they were very good practitioners of Design and Technology teaching, few of them had previously made significant use of this collaborative technique, and this alone speaks volumes of the perception of the teaching and learning regime. It is so often seen to be an individual, personal activity. One of the clearest messages from the APU data is that opening it up to group discussion in a supportive/critical framework, is a highly effective improvement strategy. We shall explore different aspects of this matter in Chapters 4, 5 and 6, using some of the case study projects from the UTA study.

Pupil differences

Whilst it is true that all the effects outlined above combine to create serious performance differences for pupils, they do not all have the same effect on all pupils. These procedural issues might be seen as things that the teacher has some control over – they are part of the 'delivery' mechanism for projects.

The variables that surround the '*receiver*' are equally significant, however, and teachers have no control here. Children come in wonderfully diverse forms, each with a unique set of qualities that make them react to the teacher's 'delivery' in equally diverse ways. It pays handsome dividends if the teacher can understand this diversity.

In the APU data analysis, we were constantly drawing attention to the different levels of performance of different sub-groups in the data.

- The crucial issue here is the performance of low-ability pupils, and the ways in which the test structure (reflective/active and procedural tightness/looseness) influence these pupils.

- A major pattern emerges in relation to the performance of low ability girls, certain test structures appearing to be far more supportive than others.

- Such is the extent of the gender imbalance in the Evaluating Products tests, that the low ability girls almost outperform the high ability boys.

- Girls tend to out-perform boys in the contexts focused on *people* (e.g. designing toys for children) . . . but there is a tendency for boys to do better in the *industrial* context (e.g. designing for production). The context of *environments* (e.g. designing Post Office leaflet distribution system) would seem to be largely gender neutral.

- Generally, the more open the task, the better the girls perform. Much tighter definition appears to favour boys.

- As a general rule it would appear that the context effect is less marked for higher ability pupils.

(Kimbell *et al.*, 1991, section 15)

The examples are endless and taken together they illustrate the folly of assuming that the performance of all pupils will be raised or lowered by any given strategy. The chances are that it will help some and hinder others. We took this fact of life one stage further in our discussion of the design of test items, suggesting:

> One is led to the somewhat sinister conclusion that it would be possible – given an understanding of the nature of these effects – to design activities deliberately to favour any particular nominated group or sub-group. More positively, it would also appear to be possible to design activities that largely eliminate bias or at least balance one sort of bias with another.
>
> (Kimbell *et al.*, 1991, section 15)

In the context of pupils' projects in schools, however, we explore these issues of individualised performance in some detail through the case studies of progression in Chapter 5 and the discussion of differentiation in Chapter 6.

The challenge of progression

We have identified above a series of factors that affect the response of pupils to tasks in Design and Technology, and we have suggested that these responses are unlikely to be the same for all pupils – since all pupils are different. It is a very real challenge therefore to find ways of discussing the *progression* that pupils might make through a Design and Technology programme. Pupils start off at different points and with a unique combination of values, understandings, skills and preoccupations. On top of this multiple variance, teachers superimpose their own classroom practices which (our evidence suggests) naturally tend to favour some groups at the expense of others.

We are not in the least surprised that progression has become a major concern – even a buzz word – in the litany of schools. And we should not be under any illusion about the complexity of the challenge that it presents. Doing one thing will help one child to progress and will cause another to regress.

We have decided, therefore, that in order to understand progression, it is necessary that we look at specifics. We have therefore devoted Chapter 5 of this book to a series of cameos of pupils' performance, outlining what they did – and why – and explaining why we believe that this represented progress for the individuals concerned. We have structured the chapter around some of the major building blocks of capability – notably the 'process'-centred ones enshrined in the Design and Technology Order (DFE, 1995). Each case study has to be interpreted as an individual case which carries messages about that pupil, in that school, with that teacher, doing that project. We have, however, sought to go beyond these specifics to tease out some more general principles that inform our understanding of progression.

A noteworthy feature of Chapter 5 is that – we believe for the first time – we are presenting case study material from all four key stages using a common observation framework. There is far too little material in the public domain that relates pupil performance *across* school phases, and the evidence suggests that this has contributed to a serious lack of perception amongst teachers about the nature of teaching and learning across the whole 5–16 year spectrum. We attempt therefore, in Chapter 5, to paint a big picture of progression from age 5 to age 16. It is necessarily sketchy and therefore somewhat fragmentary, but it represents our attempt to bring some coherence to a debate of supreme importance. The Year 6–Year 7 boundary is a particular case here, and it is so marked that we have devoted Chapter 7 to exploring it.

However, the biggest problem raised by the discussion of progression, is that it presumes a set of values. It presumes that we all know and agree about the directions in which we should be making progress. *Progress towards what?* – is a critical question, and it is one which logically must precede any other debate. If we do not agree on where we are trying to get, we can expect to argue forever about whether we have got there. We are therefore forced to consider why we are doing it at all. What is the purpose of it all?

What is Design and Technology?

We are committed therefore to the detailed analysis of purpose and principle. We need to establish the reasons for studying Design and Technology and to lay out the unique qualities that the study develops in young people. The two research projects outlined above have both grappled with this problem and recognised that the first step in any project is for researchers to define their terms.

As we pointed out in our first publication at the outset of the APU project:

> The first task for the research team has been to derive a coherent and acceptable description of the activity of design & technology . . .
>
> (Kelly *et al.*, 1987)

This is a debate about first principles; about how we define and delimit the scope of our endeavour. In both of the projects we have sought to carry forward this debate about the nature of capability in technology and the contribution it can make to the

development of our young people. Neither project was ostensibly about defining this field, but neither could sensibly be pursued without first nailing its colours to the mast.

We shall do the same here, and for the same reasons. In Chapter 2 we lay out our view of technology as a human activity and subsequently, in Chapter 3, we shall outline our position on Design and Technology as a curriculum activity. Why it is there and what it uniquely contributes to children's education.

This is the richest and most critical debate, and it is the one to which we must now turn.

SUMMARY

- Technology has only very recently developed as a curriculum activity, and there remains much confusion about what it is, how it should be taught, and what benefits it offers to young people.
- This book is based on the findings of two curriculum research projects in Design and Technology.
  The Assessment of Performance Unit Project 1985–91 funded by the DES
  Understanding Technological Approaches 1992–4 funded by the ESRC
- Both of these projects have been run at the Technology Education Research Unit at Goldsmiths University of London.
- The projects have enabled us to identify a series of agenda items that need to be addressed in order for us to take forward the understanding of – and the practice of – Design and Technology.
- We see these agenda points as follows:
  1 The *context of tasks*: the setting of the task creates meaning for the task and provides starting points for action.
  2 The *hierarchy of tasks:* the level of specificity at which tasks are set has a material influence on the demands that it makes of pupils.
  3 The *structuring of activities*: the sequence of sub-tasks equally has a major impact on the difficulty and manageability of the task.
  4 These two previous issues raise the matter of

pupil autonomy in the classroom. To what extent can they – or should they – be making decisions for themselves?

5 The iteration of *action* and *reflection*: these two facets of capability need to feed off each other, so a weakness in either damages the whole.

6 *Collaborative activity*: the opportunity to use others as a sounding board for ideas.

7 These features of the activity are (largely) under the control of teachers, but there are some things that are not. The most significant here are the *individual differences* amongst pupils and this raises the problem of differentiated activity.

8 All of these concerns combine in the challenge of *progression*. We need to build programmes of activity that allow all pupils to make maximum progress.

9 But this raises the big question – *'progress towards what?'*. We must have a view of why we are all here doing the things we are trying to do. And that is the primary question to which we now turn.

CHAPTER 2

Technology and human endeavour

Introduction

In this chapter, we shall discuss the *nature of technology* as a phenomenon that lies at the heart of what it means to be human. In technology, *Homo sapiens* ('man the understander') meets *Homo faber* ('man the maker') (DES/WO 1988). But underlying this powerful liaison of mind and hand is an infinitely more powerful force. For technologists are of necessity visionary – they imagine the impossible – they project forward from what *can* be done now, to what *might* be done tomorrow. They see things that are as yet unseeable except in their own 'mind's eye'. It is this visionary quality that has dragged the human race out of the primordial mire and placed it in a position of such supremacy that we can choose from a breathtaking range of possibilities. Shall we visit the moon today or shall we utterly destroy our planet?

The Polish scientist and philosopher Jacob Bronowski, when writing his magnificent tribute to the triumphs of humankind (*The Ascent of Man*) put his finger on this central defining characteristic of and motivation for technology:

> Among the multitude of animals which scamper, fly, burrow, and swim around us, man is the only one who is not locked into his environment. His imagination, his reason, his emotional subtlety and toughness, make it possible for him not to accept the environment but to change it. And that series of inventions, by which man from age to age has remade his environment . . . I call . . . *The Ascent of Man*.
>
> (Bronowski, 1973)

Technology is essentially about satisfying human desires – for comfort, for transport, for power, for communication, for identity. It is built upon dissatisfaction; upon the tendency (some argue the compulsion) of humans constantly to seek to improve their lot. But equally it is built on the vision of those who say '. . . I can see a better way of doing that . . .' Technology is a task-centred, goal-directed activity. It is purposeful and focused. Technology makes use of a wide range of bodies of knowledge and skill, but is not defined by them, for the *raison d'être* of technology is *to create purposeful change* in the made world. Something did not exist before, but now – as a result of human design and development – it does exist. We have wheelbarrows, wallpaper, widgets, waistcoats and warships because someone (or group) decided (for one reason or another) that they would be good things to have. We can do things with them that were impossible without them. This is technology.

But technology is not just about new things. We constantly try to make our latest model of wheelbarrow (or warship) better than our competitors. This too is technology and again it is a highly focused activity. It is also intensely value laden as should be clear from the use of the word 'better'. We might mean cheaper, or stronger, or longer lasting, or shorter lasting, or less damaging to the environment, or more damaging. All these are perfectly proper objectives that might make our whatever better than yours for the purposes we have in mind.

Technology and clients

Definitions of technology that allow us to distinguish between it and other human activities, must therefore centre on this concept of purposeful change. The boundaries of technology are *not* set by our current practices and understandings in electronics or biochemistry or any other existing field of knowledge. The boundaries are defined by our human desires. These desires may arise through dissatisfaction with a current arrangement ('I really do need a better way of fixing punctures on my bike') or simply through an opportunistic response ('I can see a great way to use silicon chip technology to make singing birthday cards'). Few of us would have thought seriously about buying singing birthday cards – or even singing birthday-cake candles – but when they become available they become desirable.

So technology may be driven by the explicit desires of the user (for a better puncture-repair kit) or by the desires of the designer or the producer looking to exploit an opportunity. Probably some innovation results from a combination of these sources, where an interesting design idea meets a novel manufacturing opportunity to meet a real user need. *But in any event, any technological innovation arises from human desires and results in a change in the made world*.

Another fact of life in technological innovation is that – regardless of the *source* of the original desire – the purchaser is the ultimate decision-maker. Any given technological outcome only exists when there is an identifiable client-based need for it. This need may have been massaged by marketing experts or it may be a real fundamental need, but without it there will be no change in the made world. It matters not whether this need/desire is for Sidewinder missiles (very few clients but very wealthy ones – hence sufficient development and production money) or for cups and saucers (very many clients – hence a big market creating sufficient development and production money). In the Thatcherite 1980s this would typically have been characterised as technology being 'market-driven', but that phrase tends to disguise the fact that 'markets' are little more than collective human desires.

The Science Policy Research Unit (SPRU) at Sussex University conducted a detailed study of 'Industrial Innovation' – project SAPHO. It sought to identify and evaluate the factors which distinguished innovations which have achieved commercial success from those that have not. The first and most critical feature was – as we might properly expect from the above discussion – all about understanding the needs and priorities of the client or user.

> The clear-cut differences . . . were . . .
> 1 Successful innovators were seen to have a much better understanding of user needs. They acquire this superiority in a variety of different ways. Some may collaborate intimately with potential customers to acquire the necessary knowledge of user requirements. Others may use thorough market studies. However acquired, this imaginative understanding is one of the hallmarks of success. Conversely failures often ignored users' requirements or even disregarded their views.
>
> (SPRU, 1984)

There are some classic design failures that illustrate the point. Clive Sinclair saw an opportunity to exploit an innovative power/transmission system for his famous C5 tricycle. But without a client-based need for it, it became a very expensive white elephant.

The process of technological innovation can work from both ends: from a user desire or from a designer/manufacturer desire. But the purchaser/user/client has the whip hand. If we decide that we do not want or need the new whatever, it is doomed to failure.

Technological and non-technological change

There are of course many kinds of user-focused purposeful change going on around us all the time. Are they all technological? Or is there a further element in the definition of technology? At the next General Election there may be a change of government. If this were to happen, it would be an expression of 'user' opinion but not many of us would describe it as a technological change. We would rather call it a purposeful political change.

Technological change operates on the made world of products and product-based systems. Any 'made world' product innovation – from a wheelbarrow to a widget – is the product of technology. Political change will naturally impact upon these 'made world' changes, but this impact does not create or define the change. Politicians influence the climate within which we live, but they do not directly create our world. Technology does. The fact of technological change is manifest in the world around us; in houses, in trains, in garden centres. David Noble's description of technology sums this up very neatly. It is, he says, like

> . . . hardened history, frozen fragments of human and social endeavour . . .
>
> (Noble, 1984)

Non-technological change

On this definition, a change in the voting system itself (e.g. from 'first past the post' to 'proportional representation') would not be an example of technological change. It would presumably be intended to provide a better constitutional relationship between the electors and the elected. It might *incidentally* involve a new technological product (there are some wonderful electromechanical voting machines in the USA), but that is not the reason for the change. Re-designing the voting system is *not* intended to change the made world and therefore would not be a technological change.

Technological change

By contrast, a new adaptor that enables us to fix a hose-pipe to a tap without spraying water everywhere *would* be a technological change (and a very welcome one), since it is specifically intended to impact directly on our made world.

We recognise that this distinction has some fuzzy edges, but it is none the less important to make. And generally it is easy enough to make. We might argue about whether a new bit of software – a computer game, for example – is a product or a system

(or both). But it is easy enough to agree that it is intended to create a change in our made world. So too was 'Mr Whippy's' squirting ice cream, Mary Quant's mini-skirt, and the British Aerospace HUD (head-up display) system of aircraft instrumentation. These technological changes are clearly distinguishable from other forms of change: from *political change* in the above example of proportional representation; from *economic change* of collectivist communism to individualist markets; and from *educational change* of selective to comprehensive schooling. Each may have had an impact on the made world of products and product-based systems, but that was not the driving purpose behind them and that is why we do not see them as technological changes. As we shall see in Chapter 3, this discussion is important in defining the limits of Design and Technology as a curriculum activity.

Change and 'improvement'

The reader will have noticed that we have been careful to use the neutral word 'change' rather than the emotive one 'improve'. Technological changes are all intended to improve something for someone, but the inevitable fact of innovation is that there are winners and losers. We can get whiter than white clothes, but only by using detergent cleaners which irritate the skin of some people. We can have cars that go from 0 to 60 m.p.h. in 3.5 seconds, but only by consuming great quantities of lead-enriched hydrocarbon fuel. We can grow far more wheat by using nitrogen-rich fertilising agents in the soil, but the nitrogen leaches out and affects our rivers. It seems almost as if there is a natural law that prohibits benefits without losses.

It is easy to identify the winners – since they are usually the clients that commission or induce the technological development. With the losers it is more difficult. Sometimes they are individuals, and sometimes whole eco-systems are at stake; sometimes they emerge rapidly and sometimes the down-side does not make itself evident for years. And sometimes the winners are also the losers. The motorway speeds up our movement but only by encroaching on land that we formerly used as recreational space or farmland.

Pacey describes this conundrum very neatly through the example of the 'snowmobile'.

> Whether used for herding or for recreation, for ecologically destructive sport or to earn a living, it is the same machine.
>
> (Pacey, 1983)

The eco-design movement is an interesting manifestation of the late 20th century. As industrialised societies have increasingly sought to build a 'better' world, we have repeatedly been brought up short by the manifest failings in our efforts. Eco-design is an interesting response to the problem not just because it seeks to illuminate the grand scale of ecological matters but because of the tools that it has spawned. Environmental audit (EA) and life cycle analysis (LCA) are complex tools that have been developed – quite literally – to *account* for each decision that contributed to the design of a product.

Whilst the surface of this argument might be in terms of whether a metal or a fibre should be used for a given purpose, the irreducible bottom line of such arguments is only reached when we get to the *values* that inform the designer and the product. This is brought most sharply into focus when the retrospective EA and LCA tools are transposed into the more proactive life cycle design (LCD) tool. Here the designer is constantly forced to confront the meaning of 'better' and 'worse' not just at the technical level of the product, but also at the levels of production, marketing, use and disposal. As Layton points out:

> The technology does not have to be as it is. Other options have been available: what we encounter is the result of decisions which reflect the value judgments of those who shaped a development which was not inevitable.
>
> (Layton, 1993)

Science and technology

The thrust of the early parts of this chapter has been to assert that the driving force behind technology and innovation is a desire to change our made world. We wish to improve things for ourselves. However, this view is somewhat at odds with a traditional assumption that technology is actually driven by science. Scientists discover things and then technologists apply them to human purposes. This argument (invariably propagated by scientists) has always appeared less than convincing to us since a moment's reflection on the history of technological endeavour reveals example after example of technology leading, rather than following, science. The Chinese built firework rockets in advance of any established theory of rocket propulsion; a functioning steam engine preceded the Laws of Thermodynamics; and Bell's telephone system depended on the electrical properties of carbon which were unknown to science at the time he used it. We shall return to this matter later in the chapter.

It seems self-evident that science can provide an immensely important resource for technology, but this resource is neither a sufficient nor even a necessary condition to guarantee technological innovation, and a useful analogy can be seen in the world of crime and detection. When a police officer investigates a murder, he or she looks for someone who had the *means* (a knife) to commit the dirty deed and the *opportunity* to use it (in the rhododendron bushes at the garden party). But most of all the search is for someone who has the *motive* (personal gain or revenge). In terms of technological innovation, science can only reliably provide the first of these – the means of technological advance. As for the second, we have only to reflect on the way technology strides ahead in wartime – because of the concentrated political/economic focus – to see that it too plays its part. But it is our hunger for improvement (in whatever way one wishes to define this) that provides the driving force of motive.

Moreover if we want something badly enough, the evidence suggests that human ingenuity will find a way to do it even without the scientific means. The examples of the steam engine and the telephone illustrate that a heady combination of motive and opportunity was more than enough to overcome the difficulty of a lack of any formal scientific base for the innovations.

Science is a resource for technology – and that is all. As Layton eloquently describes it, it is a quarry to be mined rather than a cathedral to be wor-shipped in. Science, on this view, becomes 'a servant to technology, a charwoman serving technological progress' (Skolimowski, 1966).

In case this debate appears a little arcane, we would do well to remember the damage that has been inflicted on engineering education in the UK by its too close association with science, especially in university circles. In official report after official report the universities are castigated for making engineering courses into quasi-science courses.

> Complaints commonly voiced, especially by employers, are that the education of engineers is unduly scientific and theoretical.
>
> (Finniston, 1980, section 4.18)

> . . . our overseas competitors are generally superior in the formation of engineers. This deficiency to a large extent reflects the relatively restricted and narrow conception of engineering as a branch of applied science . . .
>
> (Finniston, 1980, section 4.39)

> The engineering schools have not provided an education which is sufficiently distinctive from science and physics . . . we have tended to produce second rate scientists.
>
> (Allen, 1980)

> The whole burden of developing competent development engineers or design engineers at present falls upon industry itself.
>
> (The Corfield Report, 1976, section 8.6)

The academic snobbery attached to pure study – in Layton's 'cathedral of science' – combined with the delusion that technological innovation naturally flows from such science, has been a major con-tributor to the undoing of Britain's manufacturing competitiveness. Engineering education in par-ticular has been myopically focused on the scien-tific means for innovation rather than responding to a broader vision of technological innovation. As we have attempted to illustrate, this broader vision must incorporate the *motives* of human need and aspiration that are the real engines of techno-logical advance.

The language of technological change

A recent discussion of Neanderthal Man was prompted by the find – in southern Spain – of a Neanderthal site that is much later than others pre-viously known. The discussion centred on what qualities were possessed by later inhabitants that allowed them to dominate – and ultimately elimi-nate – the Neanderthal. The consensus of opinion was that it was not to do with muscle power or even with the overall size of the brain. Rather it centred on the development of a particular facility within the brain that resulted in superior communication skills and hence an enhanced ability to organise and to cooperate.

Communication, and especially the link between language and learning, has a well-established place in our understanding of education. The philoso-pher Susan Langer described it in the following terms.

> Language is our prime instrument of conceptual expression. The things we can say are in effect the things we can think. Words are the terms of our thinking as well as the terms in which we present our thoughts, because they present the objects of thought to the thinker himself. Before language communicates ideas it gives them form, makes them clear and in fact makes them what they are. Whatever has a name is an object of thought.
>
> (Langer, 1962)

Vygotsky makes a similar point

> Reflection . . . may be regarded as inner argu-mentation. We must also mention speech which is originally a means of communication with the surrounding people and only later, in the form of inner speech, a means of thinking . . .
>
> (Vygotsky, 1966)

For the purposes of this discussion, however, we should remember that languages are not only based on words; indeed, Vygotsky's analysis of signs and symbols as communicators takes us well beyond Langer's single-minded preoccupation with words.

In the technological world, it is these other forms of communication that become pre-eminent, but they operate in exactly the same way as Langer indicates.

> The history of engineering drawing demonstrates that the modelling methods available to designers do directly affect the thoughts they can think. Engineering drawing was a dramatic and powerful modelling tool that made possible a new relationship between management and manufacture and separated the process of design from the process of construction. It was at the heart of the industrial revolution and the new work relationships that it brought into being.
>
> (Baynes, 1992)

Whilst this is true of engineering drawing, it is equally true of all the other communication techniques used in technology. Mathematical symbols and formulae enable us to think about things that would otherwise be impossible to conceptualise. Concrete modelling enables us to think about form and structure in ways that are impossible with the use of abstract formal representations. Sketching allows us to 'talk through' ideas with ourselves or with others. Workshop manuals are full of diagrams, drawings and photographs not because the users cannot read, but because the language of images is so much richer than the language of words when one has to deal in technological matters.

Imagine the task of describing to someone how a door lock works. Imagine doing this without a door lock to point to and by using only words. It would be far more difficult than if one were also able to use diagrams. We need to be able to 'image' in our minds what is happening on the inside of the lock, so using images as the means of communication is far more efficient.

In the technological domain we might rephrase Langer (with apologies) as follows:

> Images are our prime instrument of technological expression. The things we can draw are in effect the things we can think. Models are the terms of our thinking as well as the terms in which we present our thoughts, because they present the objects of thought to the thinker himself. Before a drawing communicates ideas it

gives them form, makes them clear and in fact makes them what they are.

Those who might doubt the validity of this transformation should read *The Art of the Engineer* (Baynes and Pugh, 1981) if only for the fascinating account of the codification of engineering drawing and its central role in making possible the development of the steam engine.

> James Watt's personality, education and situation meant he was well fitted to codify drawing practice . . . he drew together the threads of architectural, technical, scientific, military and naval draughtsmanship to turn them into an effective means for design, development and production control . . . Watt executed all the drawings himself . . . step by step there evolved, from Watt's work alone, small groups of draughtsmen and finally well organised, recognisably modern drawing offices . . . [which became] the normal means for developing ideas and controlling production.
>
> (Baynes and Pugh, 1981)

The language of technology is indisputably a concrete one – of images, symbols and models. Without this language it is just not possible to conceive of technological solutions.

Expressing ideas and developing ideas

The thrust of our argument above is that the use of a concrete language is essential to grapple with the concrete realities of technological innovation. But we need to go further than this – to examine how we *use* the language as a technological development tool.

One of our principal conclusions from the APU project (see Chapter 1) concerned this critical relationship between the concrete *expression of ideas* and the *development of ideas*.

> . . . the act of expression pushes ideas forward. By the same token, the additional clarity that this throws on the idea enables the originator to think more deeply about it, which further extends the possibilities in the idea. Concrete expression (by whatever means) is therefore not

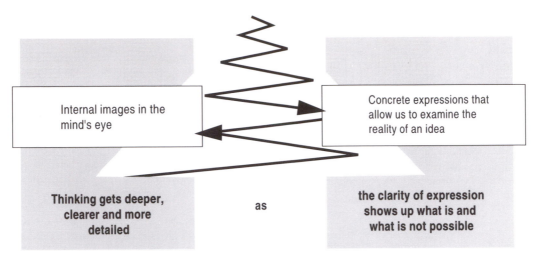

Fig. 2.1 Interaction of internal images and expressions in drawings

merely something that allows us to see the designer's ideas, it is something without which the designer is unable to be clear what the ideas are.

(Kimbell *et al.*, 1991)

Whilst this work for the APU was based on 'novice' technologists at school, an exactly parallel finding emerged at about the same time from the study of expert technologists – in this case Edison and Bell's work in developing the telephone – which was analysed in detail by Gorman and Carlson (1990) at the University of Virginia.

... the innovation process is much better characterised as a recursive activity is which inventors move back and forth between ideas and objects. Inventors may start with one mental model and modify it after experimentation with different mechanical representations, or they may start out with several mechanical representations and gradually shape a mental model. In both cases, the essence of invention seems to be the dynamic interplay of mental models with mechanical representations.

(Gorman and Carlson, 1990)

Gorman and Carlson's findings accord exactly with our own, which we summarise in Fig. 2.1. We used

the term 'iterative' to describe the interaction of the internal (mind's eye) images and their expression in drawings and models. Gorman and Carlson use a different term; 'recursive', but the point remains the same.

Our work for APU led us to the conviction that

Cognitive modelling by itself – manipulating ideas purely in the mind's eye – has severe limitations when it comes to complex ideas and patterns. It is through externalised modelling techniques that such complex ideas can be expressed and clarified . . .

It is our contention that this inter-relationship between modelling ideas in the mind and modelling ideas in reality is the cornerstone of capability in design & technology. It is best described as 'thought in action'.

(Kimbell *et al.*, 1991)

Technological capability

As with all languages, its existence is no guarantee that we can all use it fluently. The fact that we have at our disposal a complex concrete language for technological developments does not guarantee that we can all be creative with it. We can use languages at many levels of sophistication and an important distinction is between using it *to understand and respond* and using it *to create new*.

The argument is as follows: When I am in France on holiday I am content to understand and respond in the language; to use it to buy goods in the market or to converse with local people over a drink in the bar. I do not aspire to explore the limits of the language by creating something new; a novel, poem or play. So too with technology we might just wish to use the language to understand and respond: to read the symbolic instructions in the car manual when we break down; or to understand and interpret the dress pattern in order that we can adequately follow its instructions. But equally we might wish to be more proactive with the language – using it to create new technological ideas and outcomes.

This is what we take to be the essential difference between *technological understanding* and *technological capability*. By 'capability' we mean that combination of ability and motivation that transcends understanding and enables creative development. It provides the bridge between what is and what might be. Specifically, in technological terms, it mediates between human desires and dissatisfactions on one hand and technical constraints and possibilities on the other.

Technological capability is a central concern of this book. Not technological capability in general – but technological capability as a part of education. Technology has only recently emerged into the mainstream of the school curriculum. Thirty years ago it was an idealistic dream of a handful of pioneers, and even 15 years ago it was a minority activity bearing little resemblance to its current formation in the curriculum. At the heart of this new technology curriculum lies this creative technological capability that we seek to develop in all pupils.

In Chapter 3, we shall turn our attention to the educational significance of technology, specifically to explore what it uniquely contributes to the education of young people.

SUMMARY

- Technology is a quintessentially human activity based on our desire constantly to improve our condition. It is an organised way of creating purposeful change.
- It is therefore centred on, and defined by, the needs/wants/desires of people, who will be the users of the outcomes of any given technological development. Technology is task-centred.
- But not all purposeful change is technological. To be technological, the change needs to impact on the 'made world' of products and product-based systems. Warships and widgets are technological but voting systems and new economic models are not.
- Whilst technology seeks 'improvement' (for someone) it invariably creates winners and losers. It is intensely value laden and usually driven by the values of those that commission the task.
- Technology can be resourced by any body of knowledge – but is not defined by them. Whilst a rich knowledge base is a valuable resource for technology, the possession of any body of knowledge is neither a sufficient nor even a necessary condition of technological development.
- The language of technology is dominantly a concrete one; of graphics, symbols and models. The development of technology has required the progressive sophistication of these concrete languages.
- There is a tight relationship between the expression of ideas and the development of ideas. It can be described as 'thought in action'.
- Technological capability is dependent upon a combination of abilities and motivations that empower us to create what has never formerly been seen. It enables us to bridge the gap between human aspiration and technical constraint.
- Technological capability is now at the heart of the school curriculum in the UK and we must examine why it is there and what it contributes to the education of young people.

CHAPTER 3

Design and Technology in schools

Introduction

It is by no means self-evident that simply because something exists in the world it should also feature in the school curriculum. The case for a curriculum activity must rest on the idea that by studying it, young people are enriched in particular ways, and we would like to think that a 'balanced' curriculum provides a balanced enrichment. It follows from this that anyone seeking to make a case for a new curriculum activity must be clear about what it brings to improve the balance of the child's learning diet. We must be clear not only about what it brings to the curriculum – but more importantly we must be clear about what it *uniquely* brings. We shall examine therefore the reasons why technology has come to have a special place in the curriculum and the educational justifications for technological study.

From the outset we have to recognise that we are dealing with an international development. Technology is now an established part of the school curriculum throughout the world. It is generally accepted, however, that the developments of the 1970s and 1980s in the UK were very influential both on the formation of technology elsewhere and on its widespread adoption. The model of technological capability that has now been endorsed by UNESCO is recognised as owing a great deal to developments in the UK. One difference, however, is in the title. Whilst in many parts of the world 'Technology' is the established title, in the UK the subject goes by the name 'Design and

Technology'. Accordingly, we use this title hereafter whenever we refer to the curriculum activity as currently defined in the National Curriculum (DFE, 1995).

Historically, there have been a number of interpretations of technology for schools that have ranged between seeing it as a social study and seeing it as a science. Equally its justifications have varied between the purely utilitarian/vocational and the broader educational. It might be useful to examine these contrasted visions of technology in the curriculum in order that we may better understand the real reason that it now resides at the heart of the UK national curriculum.

Awareness of technology – a social study

Courses that can be characterised as developing an *awareness* of technology have typically centred on a historical, sociological, economic, even philosophical study of the *impact* of technology over given periods of history. How (for example) did the development of bridge technology affect the placement and development of communities? Why (for example) did the evolution of iron smelting take place where it did and how did it influence the development of the Industrial Revolution? Such studies are typically backward-looking – reflecting upon what has happened in the past. The more adventurous courses of this kind seek to project forwards to *what might be* the consequences (for example) of global telecommunications, or non-intrusive surgery, or cars that can steer themselves from cables buried in the road. There is a massive literature to support such approaches, and some very good materials have been developed specifically into course materials for schools. This style of technology education is best exemplified by the Science, Technology and Society (STS) movement in the USA whose espoused aim is to develop thoughtful citizens who appreciate the inter-relationships between the technological world and the social/cultural world. This type of programme has been given another spin in recent years by the 'intermediate technology' movement in developing countries. Under this banner, technological solutions (e.g. to starvation or deforestation) are evaluated very carefully in terms of their social impact.

These kinds of study do not require in-depth technological treatment as their priority is to contextualise technology into wider issues of human society. A recent NASTS conference in Washington included, for example, the following presentations:

- An analysis of the relationships between technology and government support in key sectors of global competition, e.g. pharmaceuticals.
- The impact of public opinion and political pressure on nuclear power plant design, construction and operation.
- Killing as healing: the ethical dilemmas and psychological mechanisms of the Nazi doctor.

(NASTS, 1991)

The common feature of STS courses is that they tend to treat technology as an accumulating set of magic 'black boxes' that can do clever things. In order to get started on such a course, you have to accept that satellites can do X or that suspension bridges can do Y without really knowing (and perhaps without really caring) how. As Layton points out.

> Personal involvement in technological activities
> . . . has not been a distinguishing feature of the
> STS programmes.

(Layton, 1993)

Competence in technology – an applied science

This kind of course is the absolute opposite of the 'awareness' course. It is essentially a 'hands-on' course that is not only designed to get into the nitty gritty business of how things work, but moreover is designed to enable students to become competent in making them work. And there are broadly two classes of such courses: those that develop competence through theoretical understanding and those that develop it through practical 'doing'.

In either case, the focus on developing competence almost inevitably results in specialised course structures. You can learn about – and become competent in – electronics, or mechanics, or lasers, or pneumatics or integrated circuits (in the theoretical strand) or about wood, fabrics and food (on the practical front). Because of the depth of understanding and experience that is typically involved in such courses, there is not time to span very far across the field of technological endeavour. Courses are specialised and are almost always modular, with separate modules in a range of specific technological fields. In the UK, the clearest manifestation of this style of technology education was seen in the Modular Technology courses of the 1970s and 1980s designed for 15–16-year-old pupils. More recently they have reappeared in a slightly different form as the Technology Enhancement Programme organised through the Engineering Council.

If 'awareness' courses are suspect in their grip on technological content, then 'competence' courses suffer from the reciprocal difficulty. They tend to take for granted that technology is simply 'there' to be engaged in and – almost by extension – is inevitably a good thing. Technology is seen less as a powerful influence on the nature of society and seen more as a branch of applied science or an extension of a craft trade. It is not something that is susceptible to philosophical questioning – rather it is something to get stuck into.

In these two schools of technological thought (awareness and competence) there is an obvious liaison to contrasted sectors of the curriculum. Awareness courses are the domain of the humanitarians who have no particular expertise in technology, nor any desire to acquire it. Such courses are essentially part of a *social study* raising humanitarian, ethical, political, ecological and anthropological questions. It is for the thinkers and the policymakers who do not wish to get their fingernails dirty.

By contrast, the competence course lies at the interface of the physical sciences with the craft workshop environment. It is essentially a *technical study* of materials, energy, control systems and related fields. It does not so much raise questions about these things; rather it provides answers through them. Students on such courses are comfortable with 'how?' questions , but typically very unsettled by 'why?' questions. A further – critical – distinction of these two kinds of technology lies in their perceived function in the curriculum. Typically the former would be justified as part of a liberal, general education whilst the latter would be regarded as dominantly vocational.

It was against this backdrop that a quite new concept emerged – the *capability course* – that is designed not to replicate either of these former approaches, but rather to blend them into a quite new formation. This new phenomenon is the outcome of three decades of development in the UK. It is not a passive intellectual study like the awareness courses, but neither is it a narrowly focused competence course.

Capability in technology

As we pointed out earlier, by 'capability' we mean that combination of ability and motivation that transcends understanding and enables creative development. It provides the bridge between what is and what might be. And as a basis for school programmes, it combines intellectual, practical and emotional qualities in a quite unique way.

The concept of 'capability' in technology was fully articulated in the Interim Report (1988) of the National Curriculum Design and Technology Working Group. In a profound introductory chapter to this report, the following definition of 'design and technological capability' was offered.

Our view of design and technological capability is that, at the very least, it covers all of the following:
i.   pupils are able to use existing artefacts and systems effectively
ii. pupils are able to make critical appraisals of the personal, social, economic and environmental implications of artefacts and systems
iii. pupils are able to improve and extend the uses of existing artefacts and systems
iv. pupils are able to design, make and appraise new artefacts and systems
v.   pupils are able to diagnose and rectify faults in artefacts and systems

(DES/WO, 1988)

This is a formidable list of requirements. It requires (especially in (ii)) the breadth of understanding and social concern typical in the awareness courses, coupled (especially in (i) and (v)) with the depth of knowledge and skill that is more typical of the restricted scope of the competence courses. At its heart, however, lies the essence of capability (in (iii) and (iv)) to identify shortcomings and take creative action to 'improve' things.

This view of capability in technology, however, did not just emerge from nowhere in National Curriculum committee rooms. It represents the latest phase of development of a story that is at least 30 years in the making (Penfold, 1988; Kimbell, 1991). And at the heart of the development lies a fundamental shift of emphasis from the study of *technological outcomes* (making them and understanding their social impact) to the exercise of a *technological process* (of design, development, manufacture and testing) that generates the outcomes.

We should not underestimate the massive significance of this move – particularly in the context of pupils learning in schools. It is a move from receiving 'hand-me-down' outcomes and truths to one in which we generate our own truths. The pupil is transformed from passive recipient into active participant. Not so much studying technology as *being* a technologist.

The development of this proactive process-centred view of Design and Technology may well be mirrored in other areas of the curriculum ('process' science and 'process' maths, for example). But uniquely in Design and Technology, it is the process that defines the discipline. The original National Curriculum Attainment Targets demonstrate this very neatly.

AT1   Identifying needs and opportunities
AT2   Generating a design proposal
AT3   Planning and making
AT4   Appraising

(DES/WO, 1990)

In no other National Curriculum subject did the consultations produce such a clear process-centred structure.

Given the circumstances surrounding the launch of the National Curriculum, and the significantly enhanced complexity of the definition of technology, it was perhaps inevitable that there would be considerable difficulties in its implementation. As John Eggleston pointed out:

> At the heart of the matter is the unescapable fact that technology has a far more demanding intellectual and expressive content than ever before. It requires not only complex understandings but also adaptability, initiative and creativity . . . but many children are finding the demands unattainable, and even more alarmingly so are many teachers . . .
>
> (Eggleston, 1992)

We do not intend to discuss here the traumas of those few years of National Curriculum chaos. It is a messy story of political and ideological meddling combined with a good deal of confusion and frustration in schools. The fact remains however that the 'final' evolution of National Curriculum Design and Technology[1] still leaves it as the only National Curriculum subject with a process-centred rationale, albeit somewhat simplified from the 1990 version. The attainment targets (ATs) remain procedural and focused on the activity of Designing (AT1) and Making (AT2). The concept of technological capability remains at the heart of the description:

> Design & Technology capability requires pupils to combine their designing and making skills

with knowledge and understanding in order to design and make products.

(DES/WO, 1995)

The question we now need to answer is what qualities this 'capability' approach to the study of technology is supposed to develop in young people. For it is on this claim that the justification for the study of Design and Technology in the National Curriculum must rest.

Justifying the study of technology

We should not perhaps expect too much of National Curriculum documents to help elucidate this issue, and indeed the 1990 Design and Technology Order describes the 'overall objective' of the subject to be 'to operate effectively and creatively in the made world' (DES/WO, 1990). As an overview statement, few would disagree with this, but it needs teasing out a little in order that we can see what Design and Technology offers that is unique in the curriculum.

The central curriculum claim for a capability view of technology rests in the uniqueness of its language (outlined in Chapter 2) and consequently the opportunities it presents *for exercising unique ways of thinking about the world and for intervening constructively to change it*. The claim does not rest on the hope that the outcomes of the changes will actually improve anything. It rests rather on the fact that the *process* of trying to create change requires pupils to engage in a challenging, enriching, empowering activity.

Design and Technology as creative, concrete thinking

There is an astonishing similarity between descriptions of the process of Design and Technology and the more general processes of thought. This is so marked that there are areas where the two merge quite naturally. At the end of a very long treatise on 'Thinking', Humphreys (1951) defines it as 'what happens in experience when an organism, human or animal, meets, recognises, and solves a problem'.

'Problem solving' has, over the years, had a very full treatment from psychologists, educationists and philosophers seeking to describe and explain the processes of thought. Thomson (1959) outlines the characteristics of problem-solving behaviour; Kubie (1962) relates it to a 'cybernetics' theory of learning; Vygotsky (1962) links it to the development of language and particularly to symbolic representation; Rogoff (1990) emphasises the interpersonal nature of the process; and Dewey (1968) relates it to schools and education, 'a difficulty is an indispensable stimulus to thinking'. Normally, the argument runs, we barely think at all so long as things run smoothly for us. Habit, impulse and well-practised routine help us to drift through much work and play. It is only when the routine is disrupted by the intrusion of a difficulty that we are forced to stop drifting and think about what we are going to do. Desforges (1995) illustrates this at the conceptual level in terms of our existing (extant) mental schema being thrown into disequilibrium only by new data (observations) that do not fit into it. This creates the cognitive conflict that is essential for a subsequent reformulation of a new, more comprehensive schema.

The process is seen (broadly) as one involving problem clarification, investigation, analysis and exploration, creative attack (action), and subsequent re-evaluation (Fig. 3.1). In fact it is, in generic form, very akin to the descriptions of design processes as outlined in the early 1970s literature of Design and Technology.

Designing is – in a sense – concrete thinking, and it is no coincidence that in practice designers frequently talk of themselves as 'thinking with a pencil', a quality that we explicitly encourage in young people in Design and Technology programmes.

This realisation has been at the centre of the arguments that, since the early days of design education, have carried forward the development of Design and Technology in schools in the UK,

> From the simplest problem to the most complex, the design process is concerned with the educational problem of clarifying the thought process or 'reasoning' of the child, indeed it is an attempt to lay bare before the child his [*sic*] own thinking.
>
> (Kimbell, 1982)

through to

> The conduct of design activity is made possible by the existence in man of a distinctive capacity of mind . . . the capacity for cognitive modelling . . . [the designer] forms images 'in the mind's eye' of things and systems as they are, or as they might be. Its strength is that light can be shed on intractable problems by transforming them into terms of all sorts of schemata . . . such as drawings, diagrams, mock-ups, prototypes and of course, where appropriate, language and notation. These externalisations capture and make communicable the concepts modelled.
>
> (Archer and Roberts, 1992)

This is the crux of the matter. The special feature of technology – unlike so much of the curriculum – is that the processes of thought and decision-making are exposed to the light of day. When you go through a design folder you can see the decision-making unfolding before your eyes because the graphics, the models and the prototypes are clear concrete expressions of that thinking. There are several very significant consequences of this:

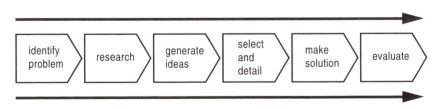

Fig. 3.1 Problem-solving process (after Schools Council Design and Craft Education Project, 1972)

- You can go back over it with the pupil to examine where critical decisions were made.
- You can look to see the basis of evidence for that decision.
- You can examine points at which alternatives would have been possible.
- You can use these as jumping off points into new lines of development.

In short, Design and Technology not only enhances the thinking and decision-making powers of young people, it also enhances their *conscious awareness* of those thought processes.[2] They not only learn to think and make decisions, they also know (and can see) that that is what they are doing.

The final twist to this tale is that, in secondary schools, the pupil constituency to which this activity has traditionally appealed is not one that had held intellectual development at the top of its agenda. Before the implementation of the National Curriculum (which made Design and Technology a compulsory study for all pupils) there was more than a fair share of difficult and disenchanted pupils in most Design and Technology classes. They traditionally found refuge and strength in the practical orientation of the learning environment. For these pupils, Design and Technology has often been a transforming experience, not just because it kept them busy and out of trouble (a rationale beloved of headteachers), but because it provides a concrete lever that can expose and get a purchase on their thought processes.

A *double bonus (generative and evaluative)*

There are two sides to the argument presented here about the uniqueness (and the beneficial consequences for teaching and learning) of the concrete language of technology. They may be summarised as the *generative* and the *evaluative*.

Any product or system that adorns (or bedevils) our world, has been through a generative process that took it from the 'mind's eye' of the designer into the reality that it eventually assumes. As we have seen, this involves the use of the concrete language of designing. And the language is being used in generative mode.

Equally, in fact more commonly, we are not the designer of the world that surrounds us so much as the recipient of design decisions made by others. We typically *buy* products rather than design and make them for ourselves to our own specification. We therefore commonly find ourselves in evaluative mode. Once again, however, the concreteness of that which we are evaluating makes it a more accessible process than if we were (for example) evaluating the literary merit of a Shakespearian sonnet. The chair tips over – the pen is uncomfortable – the keyboard is clumsy, and so on. Whilst a decision may be difficult to balance (with these pros and those cons) none the less the evidence that we use to inform the decision is often immediate concrete data. We can pick it up, feel it, look at it, try it out. Even if we require data that goes beyond that which is immediately available, we can set up some relatively rudimentary tests to gather the data that we need. For example, when '*Which*?' magazine evaluated a range of cooker grills, they set up some straightforward, concrete tests of performance, including for example,

- the area heated
- the heat-up time
- the evenness of heating
- the height adjustment
                              (Consumers Association, 1992)

It is difficult to imagine the British Literary Guild setting up such immediately accessible measures for our Shakespearian sonnet. However, the point here is not that it is *inherently* simpler to evaluate the cooker, since such an evaluation *could* go to great depths of subtlety, incorporating for example its semiotic or its anthropological significance as a cultural icon. The point is that, regardless of the depth to which such evaluations might be taken, we can establish an immediate and concrete starting point for the evaluation that gives access to all users.[3] And that is what the Consumers Association thrives upon.

In the context of teaching and learning, accessibility and the ability to make meaning carry a high premium. It is our contention that pupils will far more readily get access to meaning from concrete technological activities than they will from the far more abstract, formalised activities that permeate so much of the curriculum. Returning for a moment to Desforges' (1995) discussion of cognitive restructuring, we should remember that it is dependent upon the disequilibrium that results from the pupil *recognising* the mismatch between his/her existing mental schema and new observations. The clearer and more meaningful these observations are, the more likelihood there is that pupils will recognise the misfit if there is one. Conversely, the more shrouded and misty are those observations, the less likely it is that pupils will recognise whether there is a misfit or not. In technology systems' terms, the clearer and more immediate we can make the feedback, the better we can understand the workings (and failings) of the system.

It is for these reasons that we argue that not only does the unique language of designing lend itself to helping pupils to understand their own creative thinking and decision-making, but moreover, the clarity and concreteness of the language enables pupils far more readily to restructure their cognitive schema. In short, they will learn more readily.

Alternative justifications for technology in the curriculum

There are of course many subsidiary reasons that might be advanced for technology taking a central place in the curriculum. These arguments differ according to the view of technology that is being represented. One useful way of characterising these arguments is to see them as existing within a framework that polarises applied science from craft and educational from vocational priorities (Fig. 3.2).

In the upper right-hand quadrant lie the engineering/employment justifications for technology, espoused by the Engineering Council and many politicians and lay observers of technology who intuitively feel that technology must be in the curriculum for extrinsic (balance of payment) reasons.

> . . . the economies of the Western Democracies are fundamentally dependent on manufacturing

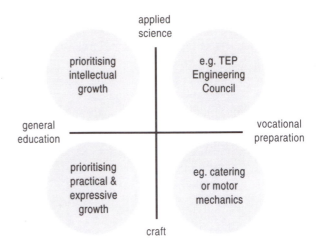

Fig. 3.2 Polarisation

> industry which is itself based increasingly on complex technologies . . .
>
> (Lynch, 1993)

Similar vocational arguments, though with a more practical flavour, are advanced for other kinds of technology represented in the lower right-hand quadrant.

> Industry needs people who are good at making things and selling them. This is why, despite the expense, practical courses must have experienced people evaluating practical capability. The epitome of this is the NVQ assessment in the workplace which are related to job-specific tasks.
>
> (Lynch, 1993)

Whilst there is some virtue in these vocational arguments for technology, one of the problems with them is that, in post-industrial Britain, it is an argument that will only be seen to apply to a small proportion of the population. Eggleston's cutting observation, that technology might provide a good vocational education for unemployment, sums it up neatly. Moreover, whilst good arguments might be presented for a vocationally related form of technology post-16 or even post-14, they have no relevance whatever to the vast majority of the school curriculum, i.e. that pursued between the years of 5 and 14. The vocational argument is further weakened by its selective vision. If the UK curriculum were really concerned with vocational preparation, we would be concentrating on vocational Maths, Science and English as well as Technology.

In the left-hand quadrants (Fig. 3.2) reside more general educational arguments for technology. The lower quadrant prioritises the visual, tactile, emotional and cultural development that stems from thorough immersion in the practice of any number of crafts. The upper quadrant prioritises justifications based on intellectual growth through the rigours and disciplines of technological study:

> The ability to identify and control variables, to engage in quantitative modes of working, and to systematically experiment to optimise the performance of a device are all skills which can be borrowed from science and brought to bear fruitfully on technological activities.
>
> (Layton, 1993)

We could have used different axes in the diagram and thereby polarised different kinds of argument for the study of technology. They are many and various. And they are all, in our view, secondary to the principal justification which is that the creative and disciplined use of the concrete language of Design and Technology provides unique ways of thinking about the world and enables pupils constructively to intervene to change it.

Making it happen in *every* classroom – the challenge of the National Curriculum

Design and Technology as a curriculum activity has grown from practice rather than from theory – from teachers in the classroom trying out innovative and often idiosyncratic activities and programmes, rather than from an academic analysis of a field of knowledge. And it has been hugely successful. Pupils voted with their feet; courses expanded and proliferated; competitions and prizes led to high-profile public exposure – where politicians and others were delighted to shake a few hands for the camera. And with the growth of Advanced level work, eventually even the universities caught up with the fact that there were some quite exceptional young talents coming through this route and increasingly sought them out.

The success that catapulted Design and Technology into the political limelight in a mere 25 years or so was built on a largely optional activity that teachers could choose to do (or not do) according to their taste. The subject was developed in those early years by enthusiasts who were committed to the activity and often passionately interested in making it work and helping it to develop. The success was therefore selective; limited to the classrooms of enthusiasts. In those schools where the teachers understood the new opportunities on offer within Design and Technology, there were quite outstanding things going on, but in many more schools, where there was less understanding and enthusiasm, there was little that would commend itself as being good practice in the developing discipline. Local education authority (LEA) advisory teams and HMI[4] strived throughout the 1980s to disseminate good practice, but it was an enormous task. Any LEA adviser of the period, if asked to identify really good practice in Design and Technology in their LEA, would typically list schools on the fingers of one hand. Even in 1988–9 when the APU national survey was conducted, the differences in performance *between schools* was far higher than it was for Science or Maths or English.[5]

In the late 1980s, Design and Technology moved remorselessly towards the centre of the rapidly emerging National Curriculum, on the strength of what can happen in established centres of excellence. But there was no strategy to deal with the much greater number of schools which were (at best) very unsure of what this new beast amounted to. As we pointed out at the time

> Someone was going to have to decide what it was that we were all doing so successfully. It was going to be written down in attainment targets (ATs) and programmes of study (PoS) – and *everybody* was going to have to do it. It had to be tamed and institutionalised.
>
> (Kimbell, 1992)

We need to remember that this was all occurring at precisely the same time that the advisory and support services and specialist centres of many LEAs were collapsing. The introduction of local financial management, which required LEAs to devolve increasing amounts of money to the schools, had catastrophic consequences for curriculum development and support.[6] This might not have been so damaging had Design and Technology been better established and more widely understood. But it was not, and consequently in many schools, National Curriculum Design and Technology was introduced by teachers who were less than confident about what they should be teaching, or how they should be teaching it.

With little support and guidance, it is little wonder that questions of quality began to appear. HMI monitoring of the first year of implementation suggested that,

> . . . many teachers have found the D&T aspects of the Order unhelpful and difficult to understand . . . in a great many schools, pupils' progress in D&T was seriously constrained because teachers were operating at the limit of their technological capability and understanding. Although the great majority of these teachers spent many hours, often of their own time, discussing the content, delivery and assessment of national curriculum technology, they had insufficient time or opportunity to develop their own skills, knowledge and understanding of D&T . . . or to evaluate appropriate teaching and learning strategies and consider levels of attainment.
>
> (HMI, 1992)

By the following year little had improved,

> . . . when judged against the broad range of requirements and the Levels of attainment in the National curriculum, much of the D&T work was undemanding and standards were low . . . many teachers were uncertain about what constitutes D&T; they were unfamiliar with the Order and found it difficult to interpret and implement; . . . in many schools planning for each Key Stage lacked coherence . . .
>
> (OFSTED, 1993 a)

This, combined with other national criticism and a third (equally downbeat) HMI report (OFSTED, 1993 (b)), prompted the National Curriculum Council to review the Design and Technology Order. The review process was overtaken by Dearing's review of the whole National Curriculum which led to the revised curriculum (DFE, 1995) that took effect from September 1995.

Throughout this troubled time, the problems that have most preoccupied teachers have been operational problems of the classroom. How to teach it. And it is these questions of practice that we are seeking to illuminate here. Chapters 4 and 5 take different perspectives on the problem – looking respectively at the role of *tasks* (how they work and how teachers can use them) and at the kinds of *progression* that might result. We do this partly through analysis and argument and partly by exemplifying it through children's work.

Retaining a perspective on why we are here

In reading these chapters however, teachers need to retain the big vision of what can be achieved through Design and Technology and what it is that sets it apart from the rest of the curriculum.

The fundamental case for Design and Technology in schools lies in two uniquely human qualities. The first is reflective, and is the ability so to focus our dissatisfactions with the world that we can pinpoint something in need of change. The second is active, and depends upon that unique capacity of the mind to manipulate the concrete language of technology so as to image and model new ideas and arrangements. These two, working in a tight iterative relationship, are the root-stock from which we can grow the technological capability of our young people. As it grows, pupils become critical without becoming disenchanted. They become independent and resourceful, empowered not just to identify weakness but to do something about it; to intervene creatively to improve things. It is an immensely satisfying capability that combines practical, intellectual and emotional challenge. It builds confidence and self esteem. And ultimately – as Bronowski points out – it builds civilisations.

SUMMARY

- There are many different interpretations of technology in schools.

- *Awareness courses* contextualise technology into the wider issues of human society.
- *Competence courses* teach specific technological knowledge and skills. They are typically specialist and modular programmes.
- *Capability courses* go beyond the aims of acquiring knowledge and skills and prioritise the *activity* of designing and making.
- Design and Technology is justified as a curriculum activity by recognising its unique concrete language (graphics and models) and the way that this enables pupils to 'see' the development of their thoughts in response to a task. This concreteness also enables technological evaluations directly to inform the development of their thinking. The immediate *accessibility* of ideas enables pupils to make meaning and to learn more readily than is the case in more formal and abstract languages.
- There are alternative justifications (often vocational justifications) for studying technology, but they are not central to the educational case of technology for all.
- In the late 1980s, Design and Technology became so successful that it was moved to the 'extended core' of the National Curriculum. But this created all sorts of problems since its success was built on enthusiasts who understood it whilst the National Curriculum requires that everyone can teach it. The move also coincided with the catastrophic collapse of LEA support services. So teachers were left without the support they needed.
- Teachers were confronted by many operational classroom problems in trying to implement National Curriculum Design and Technology and in Chapters 4 and 5 we attempt to illuminate two of the central ones – concerning the nature and purposes of *tasks* and the *progression* that pupils might make through them.
- Teachers will only be able to operationalise Design and Technology if they understand why it is there, and what it can contribute for young people that other subjects cannot contribute.
- The principal justification for the development of Design and Technology capability is that through its unique concrete language it empowers pupils to identify failings in the 'made world' and to do something about them; to intervene creatively to improve things. It is an immensely satisfying capability that encourages independence and resourcefulness and that combines practical, intellectual and emotional challenge. It builds confidence and self-esteem.

Notes

1 The 1995 Design and Technology Order sets the agreed rules for technology for the duration of the promised five-year interregnum – the period of legislative peace that schools have been promised.
2 The term 'meta-cognition' has been applied to this phenomenon.
3 An interesting parallel has been elaborated for us by Agatha Christie in her autobiographical account of her writing procedure. When grappling with the complexities of *Murder in the Vicarage*, she frequently found herself getting lost with the locations of the various characters (was Lavinia last in the library or the summer-house?). She resorted to some simple concrete modelling, using a modified doll's house and characters represented by labelled dolls.
4 The HMI Loughborough summer schools became legendary amongst those who sought to understand the new discipline and become exponents of it.
5 The APU technical term for this is 'the school effect', and it is based on the average performance of all the sampled pupils in that school. The variance *between* schools in Design and Technology was often greater than the variance *within* schools, suggesting that once Design and Technology is well established in a school it affects all the pupils, but that there were many schools where it was not established at all.
6 It is interesting to reflect on this curriculum-development function. As LEAs have collapsed, and HMI have been required to revert to their pure inspection function, and examination boards all have to fit the same nationalised standards – who is responsible for developing the curriculum? Had this situation been in place in the 1970s and 1980s, Design and Technology would never have grown to its current position. In the present regime, the only remaining independent source of expertise is in the universities, who now carry a heavy responsibility for sustaining the development.

CHAPTER 4

Tasks and activities

Introduction

In Chapter 2, we described the nature of Design and Technology as being *task-centred*, focused onto the achievement of an identifiable goal. What we now have to do is to describe the *form* of tasks and the ways in which they operate within the teaching and learning exchange. What do tasks look like, and what kinds of activity do they engender?

Over the last 25 years in the UK there have developed some characteristic approaches to the teaching of Design and Technology, and it is helpful now to reflect on these approaches in the light of our discussion in earlier chapters. Amongst other things we shall show that the recent turmoil over the form of National Curriculum Design and Technology, and the role of Attainment Target (AT)1 in particular, was quite unnecessary. The fact that it created such confusion is indicative of two things. First, the lack of confidence that teachers had in their repertoire of teaching and learning approaches. And second, the lack of a coherent professional development programme for teachers that might have prepared them for what was – by any standards – a major curriculum re-organisation.

Understanding tasks

If we refer back briefly to the discussion of technology in Chapter 2, we might reasonably conclude that the order of events surrounding a technological task might be as follows:

- A desire for change (by a potential user or manufacturer).
- One that is translated and focused into a specified task (to design and make X to do Y).
- This specification then becoming the yardstick to judge the efficacy of the outcome.

But this description of the overall process falls far short of the reality of Design and Technology *in schools* because it fails to recognise the teaching and learning dimensions that have to be built into the task. A teacher would not set pupils the task of designing children's toys because there is a shortage of such toys. On the contrary, almost every conceivable variety of toy does already exist and can be bought in the shops, and often far cheaper than it could be made in schools.

A teacher might, however, set the project *in order to teach something* to the pupils. What this 'something' is may vary from school to school, from year to year and from pupil to pupil. None the less, in education, the purpose of technological tasks goes well beyond the desire to produce outcomes (e.g. toys). The task will be devised,

- to teach skills (e.g. of working materials)
- to enrich knowledge (e.g. of electronics)
- to extend processes (e.g. of evaluation)
- to develop attitudes and values (e.g. to 'sustainable' design)
- to promote working styles (e.g. cooperation).

The 'outcome' or 'product' purposes are logically necessary. They have to be there for us to accept

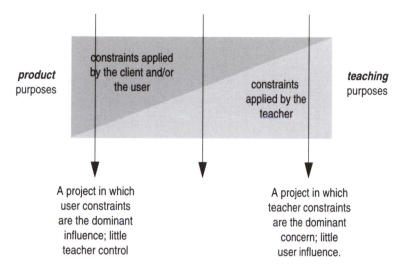

product purposes

constraints applied by the client and/or the user

constraints applied by the teacher

teaching purposes

A project in which user constraints are the dominant influence; little teacher control

A project in which teacher constraints are the dominant concern; little user influence.

Fig. 4.1 The dual purpose of tasks

that the task is a technological one, since technological tasks are about creating change in the 'made world' (see Chapter 2); they are about producing new products and product-based systems. But there are also 'teaching' purposes in the task, that transform it from being an end in itself into being a *pedagogic vehicle* for the education of the pupil.

Tasks have a dual purpose:

1 To produce an outcome that represents a change in the made world.
2 To provide teaching and learning opportunities.

There is a further duality, however, which becomes evident when we ask the question, 'Who controls the work within the task'. Again there have to be two contrasted positions. If the dominant concern is with the *product* outcome of the task, then whoever is to be the end user of the product should properly have a significant impact on how it develops. If we are designing a wheelchair ramp, we would expect to be measuring wheelchairs, talking to wheelchair users, and so on, and we should respond to their views about how the products needs to work. But if the dominant concern is to exploit the teaching and learning opportunities that are presented in the task, then the activity may have to be controlled by the teacher.

This duality might be represented in the continuum shown in Fig. 4.1, and individual projects can be located at points along it. The discussions in Chapters 2 and 3 lead us to the position where, in schools, we would require every project to have both of these two kinds of outcome. The wheelchair ramp is the *product outcome*, but the pupil's *enhanced capability* (e.g. increased knowledge and skill, or enhanced attitudes) is the other. Both are essential outcomes of the process, and we have thus derived the two central criteria by which we should judge the efficacy of technology tasks for schools.

The trick of course is to keep them in balance: too much of one and the other gets ignored – with serious consequences which can easily be illustrated through the following two examples.

Project a

A group of Key Stage 3 pupils were asked to design a brooch for Mum, to be cut from a standard size piece of copper sheet and enamelled. After the initial launch lesson, the user (supposedly 'Mum') was largely irrelevant (and was certainly never consulted) as the project centred strongly on teaching metal cutting/filing/finishing techniques and on enamelling techniques (preparation/ powder spreading/firing, etc.). When the end products were presented, the evaluations inevitably focused on the quality of workmanship (the teaching purpose) and not at all on the appropriateness of the product as a functioning piece of jewellery (the product purpose). Sharp points that would unpick clothes and puncture skin were not seen as a problem since the end product was never seriously viewed as a piece of jewellery – it was a device for teaching and for practising some specific skills. The project was enjoyed by the pupils and judged successful by the teacher. But was it a good Design and Technology project?

Project b

A group of Key Stage 2 pupils were working on a topic on bikes. One pair were looking at bikes of the future. The outcome envisaged by the teacher was a poster of new design ideas. The children initially had an idea for making a bike that had a 'turbo-power' system and at this stage the teacher attempted to support this by helping them to look at power systems. They made several drawings to explain their ideas to the teacher who attempted to get them to see the problems. They were resolute that their ideas would work, but taking them forward became increasingly difficult. From this point on they stopped consulting the teacher and became preoccupied with 'marketing' ideas, setting up contracts with other children in the class and proposing marketing gimmicks like giving away a helmet with every bike. They end up with a superficial outcome – a drawing of a bike showing no real futuristic ideas – but which they decide to market as 'the Hawk 500 . . . the bike of the future . . . for just £4000'.

These two projects illustrate the difficulty of moving too far to the ends of the continuum. The first example is arguably so far to the right that it is no longer a real Design and Technology task, since the outcome represents only the most tokenistic change to the made world. It was not seriously intended to produce a personalised piece of jewellery. The teaching purposes, and hence the

stage managing of the project by the teacher, swamped the product purposes. The other example illustrates what can so easily happen when the teaching purposes are so frail that they become almost invisible. In this event the project fails to operate as an effective learning device for pupils as they get diverted into a series of peripheral displacement activities.

The confusion of AT1

The product-teaching continuum provides teachers with an immensely difficult balancing act to maintain, and in the first few years following the publication of the 1990 Design and Technology Order, it was the source of much confusion. At the heart of the matter lay the original Attainment Target 1.

> Through exploration and investigation of a range of contexts (home; school; recreation; community; business and industry) pupils should be able to identify and state clearly needs and opportunities for design & technological activities.
>
> (DES/WO, 1990)

There are two aspects of this issue, the first of which is that tasks are seen here to derive from *contexts*. There was nothing desperately new or threatening about this. The idea of setting tasks within a context that helps to make them meaningful has in recent years become a prerequisite of good teaching. In our early work developing assessment instruments for the APU project (see Chapter 1), we expressed this concern as follows;

> The task must be set in a familiar context that then gives meaning and relevance both to that task and to the appropriate knowledge and skills that people might need to deploy or acquire in tackling it.
>
> (Kelly *et al.*, 1987)

In the following year, the Interim Report of the National Curriculum Design and Technology Working Group leant further weight to this argument;

> . . . pupils will need to bring together and use knowledge, skills, value judgments and personal qualities, the particular components and combinations

being determined by the context and nature of the undertaking . . . as the range of contexts in which D&T activity is embedded becomes broader, so the demands will expand progressively.

(DES/WO, 1988)

The essence of this move towards contextualising tasks is (as we pointed out in Chapter 1) that 'real' tasks do not and cannot exist in a vacuum, and the *setting* of the task is a major determinant of the *meaning* of that task. If you were invited to 'design a table lamp' the task would have very little meaning until you could see the context for which it is intended. It might be for one's own domestic sitting room, or for a jeweller's work-table, or for a reading room in the local library. In each case the issues that the designer needs to consider are to a large degree defined by the *context*. Equally, the success of the outcome can only be determined by examining its operation in the same *context*.

None of this would have been threatening to teachers. It is little more than common sense.

It was the second half of the attainment target that really put the cat among the pigeons: '. . . pupils should be able to identify and state clearly needs and opportunities for design & technological activities'. Were pupils really being expected to identify *their own* starting points for designing; identify *their own* need that might be met? And if so, what was the teacher supposed to do other than preside frenetically over the chaos (anarchy?) of a studio/workshop in which every pupil is doing something different? How, in this situation, would teachers ever manage to construct a teaching programme that showed any kind of progression? The wording of AT1 appeared to require teachers to move to the far left-hand end of the continuum in Fig. 4.1 and yet our criteria for Design and Technology tasks in schools, requires the *teacher* to be able to control the agenda, introducing certain things at certain times. If pupils are busily setting *their own* agendas (in response to the imperative in AT1), to what extent can teachers be said to be teaching?

Some very unfortunate activities emerged from the resulting confusion. As HMI reported in 1992,

In some schools . . . pupils often spent much unproductive time trying to identify needs; the outcomes were rarely satisfactory, and pupils sometimes became despondent about their lack of progress . . .

(DES/WO, 1992)

A problem of presentation

In reality, the problem stemmed from a lack of preparation of the teachers for what appeared to be a radical development. Few questioned the desirability of basing tasks in real contexts, but typically the only experience secondary teachers had of allowing pupils to identify their own needs for projects was with 16-year olds in GCSE projects. There was no reason to believe that this approach would work for all ages of pupils. And it did not. The problem here was one of teachers *apparently* being required to abdicate control of task setting. As we shall see, this was not necessary and certainly not desirable, but to see why, we have to pick up another strand of the story.

We outlined in Chapter 1 the notion of the *hierarchy of tasks* which has, at one extreme, a very open and ill defined context and, at the other, a highly specified task (Fig. 4.2)

As we pointed out in Chapter 1, this was a model that we used for deriving test tasks for the APU survey in 1988–9 and about which we were subsequently able to comment in terms of the effects of these different levels of specificity on different sub-groups of pupils (e.g. girls as against boys). It matters little how many steps exist in this hierarchy, but it is important that we see the ever more specific tasks deriving originally from the context and subsequently through each successive layer. The hierarchy is from the generalised context to the increasingly particularised task. (Fig. 4.3)

This progression – from the general to the particular – is nothing more than a recognition that all particular tasks exist somewhere in more generalised contexts. *But it does not follow that particular tasks must always be derived through the same progression from the general context.*

The fact that the hierarchy exists does not mean that it has to be operated in the same way each

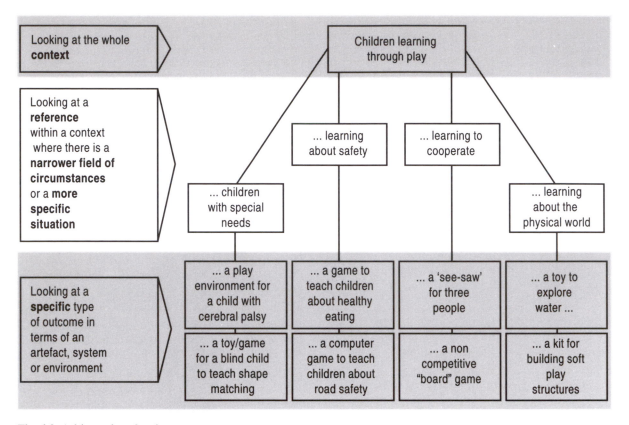

Fig. 4.2 A hierarchy of tasks

time when deriving tasks from within it. It is quite possible for example to imagine two quite different modes of operation (Fig. 4.4)

The teacher might wish pupils to get involved in designing with textiles, and may set the task of designing *a travellers' body purse* to enable the user to carry around money or other valuables whilst on holiday. This specific task exists in its hierarchy in which the overriding context might be 'protection' and which might include the layers indicated in Fig. 4.5.

There may of course be any number of layers in this hierarchy, but the point here is that it is perfectly possible to *start* with the specific task. The

Fig. 4.3 Particular tasks can be derived from general contexts

Fig. 4.4 Particular tasks can give access to broad contextual issues

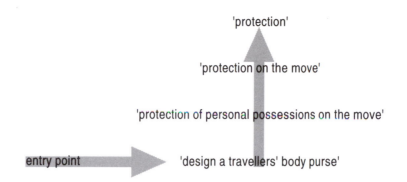

Fig. 4.5 The task is only the *entry point* to the hierarchy

significance of the hierarchy is that we would thereafter expect pupils to explore up and down it in order to inform their design of the body purse. They would need, for example, to examine the kinds of personal possessions to be protected, and how this is affected by being on the move, and even ultimately what it might mean to 'protect' them. It might mean *hide* them or *disguise* them or *fix* them. *The task is only the entry point to the hierarchy*. We might wish pupils to enter at the specific level or at the general level. But wherever they enter it, they will be exploring contexts to identify the needs and opportunities involved in their task. They would therefore be conforming to the requirements of AT1 as outlined in the 1990 Order.

Contexts are nothing more than a vehicle to bring design issues into the open. By recognising that any particularised task sits ultimately in a more generalised context, we encourage children to see where they might investigate to illuminate their designing. Furthermore, provided they always see this continuum of the general to the particular, it does not matter much where they start. Except of course that one would expect that across a whole Design and Technology programme, pupils would experience the full range of starting points.

Most often, tasks would not be set at such a tight level of specificity as the body purse, nor at the level of looseness of 'protection'. They would be somewhere in between. But again this would only be the entry point from which pupils should operate up and down the hierarchy. They would explore the general issues and gradually define for themselves a particularised task (Fig. 4.6).

The evidence from the APU survey, about the pupil performance that results from using these different layers as entry points, is hardly surprising. It indicates that different levels of task setting generate quite different kinds of response from pupils. Some pupils are able to cope with open tasks and are quite able to work down to a specific task that they are able to pursue. Others find it very difficult to do this and are much better at working on specifics and exploring the general through the specific.

The message for the development of Design and Technology capability is obvious. Part of the nature of technology (as we saw in Chapter 2) is the requirement to identify and focus on the needs to be addressed – the opportunity to be exploited –

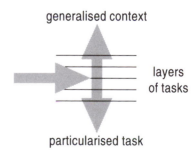

Fig. 4.6 The task entry point is typically not at either extreme

the problem to be solved. Pupils should be able to do this at all levels of specificity, since real-life problems seldom present themselves in a neatly packaged form. Learning to cope with the different levels of task is a central part of learning to become capable. It follows therefore that teachers should familiarise pupils with all these levels by – at different times – setting tasks at very different points in the hierarchy.

If teachers do this, it also tends to overcome the problem of different sub-groups preferring different styles of task. They must all get used to all styles, and this is a matter that we shall pursue in Chapter 7 when we look at differentiated performance.

Returning to the duality that we identified earlier in our analysis of tasks (the duality of teaching purposes and product purposes), it will be evident to the reader that the issue (and the confusion) of AT1 adds a further dimension, as shown in Fig. 4.8.

The initial continuum provides us with a way of looking at the question 'What is Design and Technology in schools?', but it does not address the question of how to do it. The sad story of the widespread misinterpretation of AT1 has raised awareness of this further set of issues that might be represented as follows. The critical question here is 'How do we design tasks?' (Fig. 4.8).

It is quite possible, therefore, to behave responsibly as a teacher whilst at the same time – in

different projects – requiring pupils to operate right along this continuum. Sometimes the teacher will choose to initiate activity under conditions that provide very tight constraints. At other times it will be appropriate to allow sufficient elbow-room to enable pupils to grapple with the challenge of pinning it down for themselves. Both will be expected to provide different challenges for pupils, and this diversity of experiences is necessary for pupils to develop their capability. Furthermore, both were quite allowable and appropriate under the regime outlined in the original AT1. It was not until Level 7 (GCSE equivalent) that the original programme of study included *open-ended exploration* of contexts in order to derive *their own* tasks:

> . . . pupils should develop activities which offer opportunities for open-ended research leading to the identification of their own task . . .
>
> (DES/WO, 1990)

At this level, with an established tradition of pupils at age 15 or 16 working on their GCSE major projects, few Design and Technology teachers would have had any difficulty with the idea. It is greatly to be regretted that the misinterpretation of AT1 at lower levels has resulted in its subsequent elimination from the current Order. We have removed a strand of activity that is vital to the development of pupil capability. It might have been different if teachers had initially been presented with the

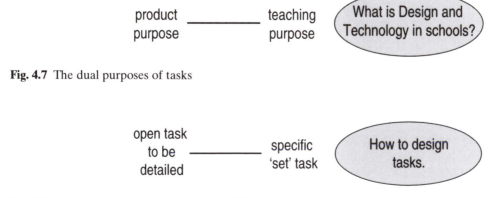

Fig. 4.7 The dual purposes of tasks

Fig. 4.8 Tasks exist on a continuum of specificity

notion of task hierarchy, so that they could develop ways of weaning pupils towards the autonomous derivation of tasks.

There are important messages here for the interpretation of the new (1995) Design and Technology Order. We are now presented with three kinds of task – two of which are 'Design & Make Assignments' (DMAs) and Focused Practical Tasks', and already one reads schemes in which the former are seen as independent project work by pupils – and the latter are seen as instructional devices for teachers. This is desperately depressing, since once again it fails to see the continuum that connects the two. They should not be seen as different things, but rather as tasks existing at different ends of a continuum. 'Focused' tasks merge imperceptibly into DMAs as the focus gets progressively less sharp. It would be very dangerous for the future development of Design and Technology if these two categories of tasks were seen as separate and that teachers completely control one and pupils the other. It is the messy middle ground that is critical to pupil's learning. Through negotiation and discussion with teachers, pupils must progressively learn for themselves the art of deriving sensible and manageable tasks. They will only learn to do it if they are given the opportunity to do it. We shall explore these issues further through examples of children's work in Chapter 5, through the discussion of differentiation in Chapter 7 and in Chapter 8.

Tasks and sub-tasks

An extension of the debate about the setting of tasks is the question of who is responsible for running it once it has been set. To what extent should teachers control the flow of the project or (conversely) how autonomous should pupils be in deciding what to do next? Design and Technology tasks never come in simple, single-issue 'bites'. They are typically multifaceted and require a degree of project planning: 'first I'll do this, then I'll be able to do that, and then it should be possible to . . . etc'. In short, tasks break down into sub-tasks.

One of the more obvious objects of education is to develop the ability of pupils to manage themselves; to bring them to the point where they not only understand what it means to take responsibility for their actions, but moreover they have expertise in so doing. Developing pupils personal autonomy would rightly be claimed by any teacher as a central goal for education.

This issue will most frequently be seen by teachers in terms of how tightly they structure pupils' activity or – conversely – how much elbow-room they allow. Assuming a given time is available for an activity (say 8 hours), it can be managed in any number of ways. It might be structured very loosely, with large blocks of time available to the pupils working independently and only limited interventions by the teacher to steer the activity. Alternatively, it might be structured much more tightly, with more interventions and instructions by the teacher. In tightly structured activities there is plenty of support and pupils do not typically 'lose their way' in the activity. But by the same token, if teachers keep intervening, there is little opportunity for pupils to learn the skills of personal management.

The studio-workshop environment of Design and Technology, in which projects typically run over an extended period, is an ideal environment within which to develop autonomous decision-making by pupils. Within this environment, pupils need to be introduced to the magnificent breadth of what is possible with materials, tools and a progressively more bewildering array of technologies. But at the same time, we have an ideal setting within which to develop their personal decision-making and responsibility. We have long held the view that technology teachers are almost uniquely fortunate in operating within this rich framework

> . . . the child will move in small steps from almost total dependence on the teacher to almost total independence . . . The function of the teacher . . . is to to steer children towards the goal of independent thought and action along the tortuous path of guided or supported freedom.
>
> (Kimbell, 1982)

Projects would be expected gradually to place ever greater responsibility on the pupil, and accordingly the teacher's intervention might reasonably be expected to get less frequent as pupils begin to master the rules of personal management. Early projects we might expect to be tightly constrained, allowing little deviation from the parameters set by the teacher. But gradually we might expect these constraints to become negotiable and permeable to the point where GCSE projects (especially in Year 11) would be only loosely controlled by the teacher and A-level projects would be almost entirely at the discretion of the pupil, involving only tutorial dialogue with the teacher.

This was what we were expecting to find when we launched our UTA research study (see Chapter 1), monitoring the minute-by-minute experience of pupils through their tasks. As we pointed out in Chapter 1, this study is spread right across the four key stages and we have followed the experience of pupils in almost every year of the 5–16 curriculum.[1] Our findings have dispelled our comfortable illusions about the progressive development of personal autonomy.

Uncomfortable research evidence

Among the many observations built into the UTA observation schedule is one that registers the points at which the teacher is *directing* the pupil to do something in particular or is *supporting* the pupil when they are trying to do something of their own choosing. This provides us with a crude but simple way of representing the autonomy being exercised by the pupil. Theoretically, the teacher might be directing or supporting in 100 per cent of the project time,[2] but in reality this never happens. The approximate norm for the projects is that teachers 'direct' activity for about 10 per cent of the time and that they 'support' pupils doing what they want to do for a further 10 per cent of the time. This leaves (on average) about 80 per cent of the time without any direct pupil–teacher interaction.

This is something of an oversimplification of the position, however, for in reality the balance of direction and support is not constant through a project. In some cases the instruction is heavily front-loaded at the outset of the project (as pupils are trying to focus and particularise their task) and in others it is more middle-loaded (as pupils confront the making activity).

In either event, we have been struck by the consistency of one trend in the data. When we plot the 'direction' and 'support' data across the whole sample, and organise it according to years (1–11) a fascinating picture emerges. We find individual support and teacher direction in equal balance throughout the whole six years of primary schooling, and we find a quite astonishing upheaval of this steady six-year pattern at the instant that pupils transfer to secondary school (Fig. 4.9).

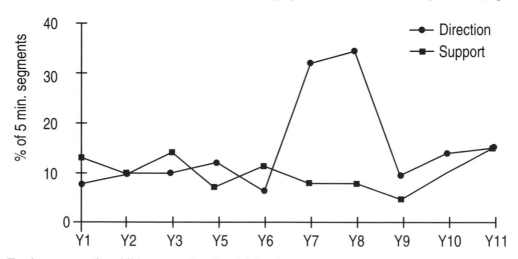

Fig. 4.9 Teachers supporting children - or directing children*

* The sample contained no Year 4 pupils

In the space of a summer holiday, pupils' perception of teaching and learning has to be completely reconstructed. Design and Technology projects have become dominantly instructional, with 35 per cent of total working time on the project spent doing as they are told. And it goes without saying that this has an enormous impact on what we had supposed to be the steady development of autonomous decision-making by pupils.

It is important to emphasise that these are average figures for all the projects in the UTA sample in each year and as such they cannot be the result of an aberration in one school. The principal conclusion that one has to draw from this is that there is something unusual about Key Stage 3. It does not fit with the experience of Key Stages 1,2 or 4 into any progressive pattern.

We shall explore this issue in detail in the following chapters. Suffice it here to observe that in terms of the growth towards personal autonomy, the Year 6–Year 7 boundary appears to represent a major step backwards. Our observation notes of the classroom confirm this. From a condition of relative independence and responsibility which is the norm for pupils in Year 6, pupils in Year 7 have reverted to a frightening level of dependency on the teacher. They wait to be told what to do, even when they know perfectly well (and are prepared to tell you) what they might sensibly do next. They seldom do it, preferring to join a queue of other similarly timid souls waiting to ask the teacher what they ought to do.

This issue brings a third continuum into the debate about tasks. We have already debated the existence, and the influence, of the first two, shown schematically in Fig. 4.7 and 4.8. The third concerns the *running* of the project, raises questions about the identification and management of the sub-tasks, and ultimately governs the level of autonomy that pupils are able to exercise (Fig. 4.10). Again this is not – or rather it *ought* not – to be a question of either/or. It should not be a matter of whether pupils or teacher see themselves as responsible for progress through the task. It should be a matter of pupils *learning to take responsibility* for themselves.

Discontinuity across the four key stages

We have been concerned in this chapter to focus the discussion about tasks in Design and Technology. We began with a debate about what they are in principle; we moved to a consideration of who is responsible for setting them, and ended by looking at how they operate (and who makes them operate) in the classroom. At each level we have argued that there should be no black or white simplistic divisions of responsibility between teachers and pupils. Rather, we have argued that it is in the messy middle ground of these continua that there lies the greatest opportunity for teaching and learning. Logic might suggest that as pupils progress through the school they might take ever greater responsibility in these matters, but the unnecessary trauma of National Curriculum AT1 and the more recent evidence that we have gathered about pupil autonomy in projects does not make comfortable reading.

Taken as a whole, our data suggest that technology tasks – and the projects that flow from them – are seen as very different things in the four key stages. When we combine the observation data with the more discursive and interpretive data derived from conversations with teachers and pupils, the different characters of Design and Technology across the key stages begins to emerge.

pupil autonomy ———— teacher control

How to run projects.

Fig. 4.10 Who takes responsibility for decision-making?

Cultural technology is characteristic of Key Stage 1. 'Technology is part of life and is all around us'. Projects tend to be topic centred across the whole curriculum (e.g. 'explorers') and technological activity derives from within the topic, involving perhaps the 'covered wagon' transport system of the early American explorers, or the means of navigation (compasses, clocks and 'shadow sextants') of the ocean voyagers.

Problem-solving technology is characteristic of Key Stage 2. 'Can you make it work?'
Projects often have a fixed starting point that then allows pupils to explore its limits, e.g. a wood strip chassis (using 'Jinks' corners) that one child explores in terms of designing an elastic band-powered and controlled vehicle, whilst another uses the chassis to develop a fairground roundabout.

Disciplinary technology is typical of Key Stage 3. 'You need to know about this (knowledge/skills)'.
Projects are designed to teach a small specified range of skill and knowledge. Pendants (to teach metal fabrication and enamelling), alarms (to teach simple circuits and sensors), rock cakes (to teach ingredient mixes and processing). The product is principally a motivational hook.

Simulated technology emerges at the interface of Key Stages 3 and 4. 'This is how real designers work'.
A gradual move to individual projects – identified by the pupils themselves and therefore generally having some reality – within which they are expected to be rigorous in the application of an abstracted designerly process and the development of a portfolio that reflects it.

The boundaries of the key stages tend to blur these distinctions and the titles are only intended to suggest broadly evolving patterns in the nature of Design and Technology activities. Nevertheless, the evidence from this study is clear. There is – at present – no such thing as a generalised Design and Technology task. Teachers have very different teaching/learning agendas in the four key stages,

and these impact so heavily on the tasks that pupils pursue, that at the moment we can only sensibly talk about a particular key stage task. In terms of practice in schools, there is little common ground between key stages that allows us to speak about Design and Technology tasks *in general*.

Yet the serious differences that we have observed and outlined here only arise because teachers place a different interpretation on the identical issues that face all Design and Technology teachers, regardless of the key stage within which they are operating. These are the issues that we have sought to elucidate in this chapter.

- The balance between the product purpose and the teaching purpose.
- The relationship between particularised tasks and generalised contexts.
- The balance between teacher control and pupil autonomy.

It would appear from our UTA study that teachers in different key stages are making very different decisions about these big issues. These differences result in what appear to be huge discontinuities in the experience of pupils. We should not be too surprised about this, since the whole activity is so new in the curriculum that there has been no time to explore the relationships between the contrasted teaching repertoires that have evolved.

Neither, however, can we afford to ignore the reality of what has emerged here. It would seem to us that it is very important for the future development of Design and Technology that teachers across the 5–16 curriculum begin to share their understandings about tasks – and the demands that they make of pupils – and the teaching repertoires that might properly be employed. We are not suggesting, of course, that all key stages should look alike. It may be quite appropriate for different key stages (or different years) to emphasise different facets in the development of capability. However, we should do it in the knowledge of what others are doing and in relation to some bigger plan.

To this end, in Chapter 5 we shall present work from pupils across this 5–16 spectrum. The first four examples are of characteristic whole projects from the four key stages, and subsequently we examine a series of critical facets of capability, illustrating how they manifest themselves at each key stage. We hope that this will contribute to a broader debate about the *development* of capability.

SUMMARY

- One of the defining characteristics of technology tasks is that they have an outcome that results in some change in the 'made world'. The wheelchair ramp changes things (or improves things?) for wheelchair users.
- But technology tasks for schools must have further qualities. They must enable the teacher to teach things. The task becomes a *pedagogic vehicle* for supporting learning.
- These two concerns need to be kept in balance in any technology task for schools and this issue was highlighted in the confusions surrounding AT1 in the 1990 National Curriculum Order for technology.
- The AT1 problem was caused by an unhelpful polarisation of the locus of control. *Particular tasks* exist on the same continuum with *generalised contexts* and it is (and always was) perfectly valid to initiate a task at either end or even the middle of the continuum. In any event, the teacher can properly retain control of task setting, but the task will be illuminated by the context. Tasks exist on a hierarchy from the general to the particular and pupils need to become capable of handling all these levels.
- This message is now vital for the interpretation of the new Design and Technology Order (DfE, 1995) since Design and Make Assignments and Focused Practical Tasks are increasingly talked of as though they were separate things, rather than as being tasks that exist at opposite ends of the continuum.

- Any task breaks down to a series of sub-tasks, and the nature of the sub-tasks has a serious impact on the difficulty that pupils will experience in doing it.
- The aim is to bring pupils to take increasing responsibility for managing themselves, i.e. for organising their own sub-tasks. The evidence suggests that this is commonplace in Key Stage 2 and almost unheard of in Key Stage 3.
- This raises the question of discontinuity across the key stages and leads us to postulate four models of Design and Technology that reflect current practice at the four key stages:

    Cultural Design and Technology
    Problem solving Design and Technology
    Disciplinary Design and Technology
    Simulated Design and Technology

- Yet the differences only arise because teachers make different decisions about the three critical issues identified in this chapter:

    1 The balance between the product purpose and the teaching purpose.
    2 The relationship between particularised tasks and generalised contexts.
    3 The balance between teacher control and pupil autonomy.

- Teaching repertoires have grown up independently in response to the different interpretations of these three issues, and increased communication of teachers across the key stages is essential if we are to share and develop a 'big picture' of progression from 5–16 years.

Notes

1 Year 4 is the only missing cohort in our sample.
2 Whilst the observations were continuous, our recording schedule was based on 5-minute segments.

CHAPTER 5

Progression towards capability

Introduction

In Chapter 4, we looked at some of the complexities in planning Design and Technology activities and, using case study data from the UTA project (Chapter 1), we outlined the different ways in which this manifests itself across the four key stages of the curriculum. It is now time to explore this area more directly, by tackling the issue of progression. We propose to do this at several levels:

- Through case study projects we shall outline some *overview characteristics* of pupil performance to exemplify what we mean by quality work at each key stage. We use this to build an overview picture of progression.
- Through case study materials we shall identify a series of *facets of performance* that we believe to be central to the development of children's capability. We shall exemplify:
  investigating
  planning
  modelling and making
  raising and tackling design issues
  evaluating
  extending knowledge and skills
  communicating.

In each case we shall (on one double page) define and exemplify each of these facets of capability and identify some key indicators of quality. We shall then (on the subsequent double page) show how this quality progresses across the four key stages.

- Finally, we shall examine some of the common issues that emerge as being particularly significant for the development of capability.

Capability across four key stages

As we have outlined in Chapter 3, we see capability in Design and Technology as going beyond *awareness* of technology and *competence* in handling knowledge and skills to '*being* a technologist', engaging in creative action, and operating both reflectively and actively in the process of designing and making. In looking at the way individual children have responded to Design and Technology challenges it has been clear from the start that some young children operate *procedurally* in a remarkably capable way; they do this within their own levels of experience which inevitably imposes limitations on the knowledge, skills and understanding they can bring to bear on the task. But, at the level of a six-year old, they demonstrate capability.

Through looking in more depth at these 'capable' children, we have sought to unpick some key issues for the progression of capability. Over the next few pages we present brief case studies of four young learners, one from each key stage. These case studies have a dual function. First they allow us to illustrate the differences and similarities of children of different ages operating effectively, and second they allow us to highlight the facets of capability that we make the focus of the subsequent part of this chapter.

Case Studies

Capability at Key Stage 1: David (Year 2)

Looking at the life of Christopher Columbus, David's class had become explorers. They set out on an ocean adventure, had been pursued by pirates and been shipwrecked on a deserted island. They found the island to be partly rainforest and there were sightings of wild animals. They decided that they needed shelters.

David worked with his friend John. They began by *investigating* books on rainforests and looking at shelters made by others. They talked through early ideas, identifying key *user issues*, such as keeping dry and safe from pirates and wild animals. Ideas were initially *modelled* through discussion and drawing 'in the air', and this was followed by them making a *plan* showing a house on stilts with a sloping roof. Different parts were indicated by a simple exploded diagram. David

worked out how to make the house section using card and joining with tabs (knowledge brought from a previous activity). His teacher helped him address certain *making issues*, measuring sections to make them fit and scoring card so that it would bend more accurately. While trying to join stilts to the base, he discovered that bracing them against the inside corners gave the strongest structure. With John, he *investigated* and *evaluated* ideas for camouflage, eventually disguising it as a tree – painting the walls brown and covering the roof with leaves. Entrance was via a drawbridge – a proposal aimed at meeting the *user issue* of security.

David was full of ideas and worked with commitment to turn these into reality. He brought both innovation and practicality to the task and made all decisions by considering a combination of *user* issues and *making* issues. However, he was working with limited, *black box understanding* (see page 74) and very limited skills in both *making* and *communicating*.

Fig. 5.1 Making the house section by joining card with tabs

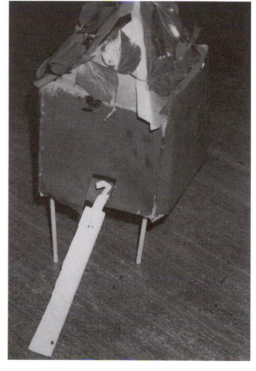

Fig. 5.2 David's finished shelter for a deserted island

Capability at Key Stage 2: Leanne (Year 6)

Leanne's class were making model powered vehicles. They discussed types of energy that could be used and were shown how to use 'Jinks' corners to make a basic frame. While other children took more straightforward routes, Leanne decided that she wanted to make a roundabout, despite everyone (including her teacher) saying this was very ambitious. Leanne rose to the challenge and started by *modelling* her initial idea, drawing it out on paper and labelling it. While she considered aesthetic *user issues*, her main focus was on *making* issues. She had ideas for using weights and pulleys to power the roundabout and *investigated* this by trial and error and by reading Lego Technic instruction sheets. She *planned* her constructing in a step-by-step chart and by completing a plan-and-review diary in each session.

Leanne engaged in extensive problem-solving to complete her roundabout. She *evaluated* through review sheets and discussion, showing herself to be resourceful and persistent as she strove between action and reflection. She generally worked with a *street level* of understanding (see page 74) although with energy systems she developed this through the project to a *working knowledge* of gears,

Fig. 5.3 Leanne working on her roundabout

pulleys and weights. She used various forms of communication such as discussion, sketching and planning charts, producing her project report with the use of Information Technology.

What am I doing to day?	What have I done?	Am I pleased?
Finishing base, put wheel base on make card top and bottom	I have done everything I said I would!	Yes
I want to make the top of the round about and my movement done horses.	I made the top and two horses plus the cogs on sticks.	Yes
I have to fix my horses on because ~~Some one~~ broke them.	Stuck one horse on and made my coggs	NO!

Fig. 5.4 Leanne's plan and review chart

Capability at Key Stage 3: Hien-Huy (Year 9)

Hien-Huy was designing and making a high-energy snack bar for his end of Key Stage 3 Design and Technology assessment. From the outset he considered *user issues* in tandem with *making issues*. For example, he thought (regarding users) about nutritional needs and wanting an easily divided bar on one hand. On the other hand he also considered (regarding making) the ingredients and ways of cooking and shaping the product. He *investigated* existing recipes and, from these and his own previous experience, produced a range of his own ideas, each of which was prototyped and *evaluated* in order that he could choose the most effective solution. Ideas were *modelled* through both drawing and *making* trial bars, and modifications were made following ongoing *evaluations*. These were carried out to criteria which allowed him to address *user issues* in some depth. *Planning* involved organising his ingredients, the method by which he proposed making the bar, and detailed time planning to ensure that he reached the required stages in the time provided.

The development of the final bar involved a high level of refining and detailing, through which Hien-Huy showed *working knowledge* (see page 75) of materials, energy systems and meeting people's needs. He demonstrated good making skills and a range of communication skills, including making notes, flow charts, sketches and labelled drawings.

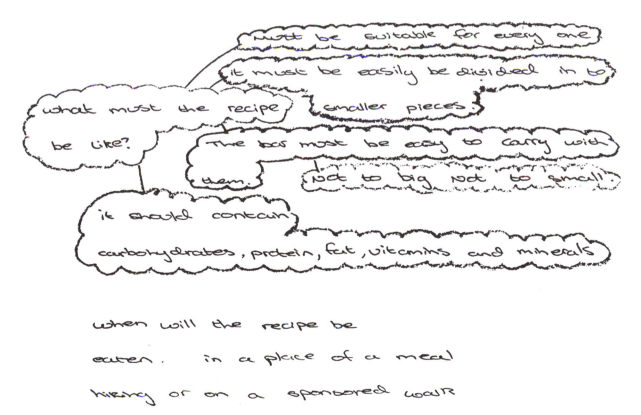

Fig. 5.5 Hien-Huy's initial brainstorm

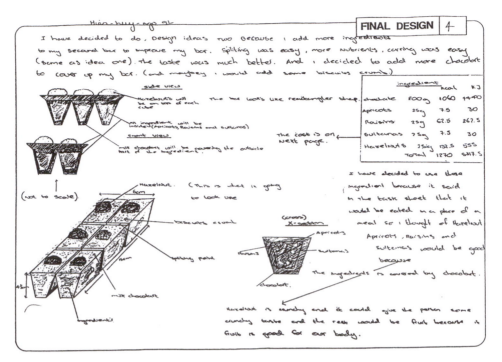

Fig. 5.6 Final ideas for the snack bar

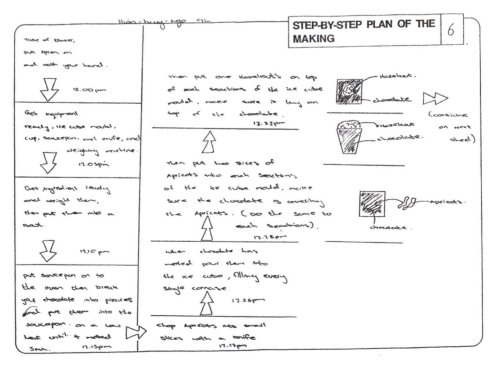

Fig. 5.7 Hein-Huy's time plan

Capability at Key Stage 4: Stephanie (Year 10)

As part of a thematic mini project, Stephanie created a brief focused on a company that wanted 'to provide a collection of fashion items based on the theme of the fairground'. She proposed a range of ideas and then developed one item (a beach bag) as a prototype. Working concurrently on ideas for the fabric and the bag she considered *user issues* such as what the bag would carry and the need for making the fabric visually exciting. She also dealt with *making issues* relating, for example, to ways of using fabrics, threads, dyes and paints to translate her fabric design ideas. She *investigated* ways of applying colour to fabric both by collecting information on techniques and through hands-on exploration. She also explored types of fastenings from books and seeing what was available in catalogues and in the shops. In *planning*, she considered ways of using her time – identifying those aspects where she needed expert help or the use of particular resources in the school. She consciously planned other work to be carried out elsewhere. Ideas were *modelled* through sketches, exploring repeat patterns, producing test samples and making paper patterns. *Evaluation* was carried out through ongoing review and reflection as ideas were tried and developed. She produced a detailed final evaluation, carried out to a school proforma, including direct reference to National Curriculum attainment targets.

Stephanie worked autonomously, drawing on previous experience and seeking out new knowledge as required. Ideas were detailed and developed through a clear process of action and reflection. She demonstrated well-developed making skills and *working knowledge* (see page 75) of materials, aesthetics and meeting people's needs. She used a range of communication techniques in a responsive way, mounting her ongoing work to produce an informative project report.

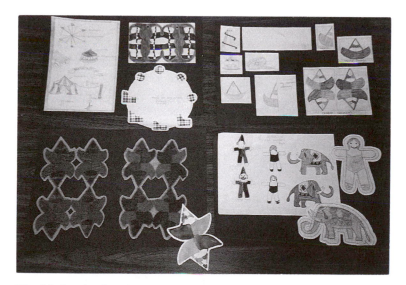

Fig. 5.8 Stephanie's developing ideas presented as a project report

Fig. 5.9 The finished beach bag

INVESTIGATING

What is investigating?

Investigating is any activity which involves pupils in collecting information which is directly relevant to their task. This could be achieved from a wide range of sources, including books, CD-ROMs, experiments with processes, materials tests, conversations with, demonstrations by, or questions to the teacher or any other 'expert'. It occurs at any point in a project.

Fig. 5.10 Key Stage 1 Investigating: using books to find out about rainforests

Fig. 5.11 Key Stage 2 Investigating: finding out together by hands on trial and error

Quality investigating includes:

- children drawing on their own experiences;

- making appropriate and relevant use of a range of reference and support materials, including books, magazines, videos, CD-ROMs, artefacts/handling collections, etc.;

- talking to their peers and look at each other's work;

- making use of group and class discussion about issues relevant to the activity;

- testing the feasibility of their ideas;

- using 'hands-on' testing of materials;

- making visits to investigate real situations;

- interviewing relevant people, e.g. clients and other 'experts'.

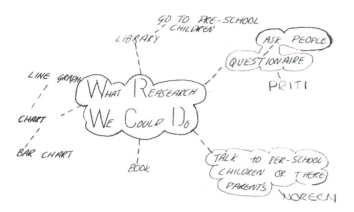

GO TO PRE-SCHOOL CHILDREN
LIBRARY
ASK PEOPLE
QUESTIONAIRE
PRITI
LINE GRAPH
WHAT REASEARCH WE COULD DO
CHART
BAR CHART
BOOK
TALK TO PRE-SCHOOL CHILDREN OR THERE PARENTS
NOREEN

Progression in investigating is supported by:

- providing a rich and varied supply of resources – including visits, outside speakers, etc.;

- providing opportunities for pupils to discuss with each other and the teacher;

- setting and structuring a task which involves the pupil in considering the needs of the user;

- helping pupils to settle onto a task with an appropriate degree of challenge (i.e. ensuring the project is demanding without being beyond the pupil's capability);

- encouraging pupils to be discerning and selective, and only investigate and record information of *direct* relevance to their project.

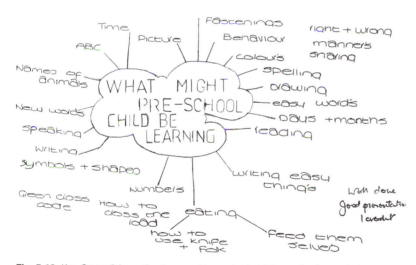

Time
Picture
Fastenings
Behaviour
right + wrong
manners
sharing
ABC
colours
Names of animals
spelling
WHAT MIGHT PRE-SCHOOL CHILD BE LEARNING
Drawing
New words
easy words
speaking
Days + months
writing
reading
symbols + shapes
Green cross code
Numbers
writing easy things
how to cross the road
eating
well done
good presentation
1 credit
how to use knife + fork
feed them selves

Fig. 5.12 *Key Stage 3 Investigating: working out what information is needed*

Fig. 5.13 *Key Stage 4 Investigating: finding out by testing*

INVESTIGATING
KEY ISSUES FOR PROGRESSION

At Key Stage 1, children need to be set tasks that will promote investigation, both of the familiar and of new areas with which they can justifiably engage. They need to be taught ways of finding out information, possibly drawing on work from other curriculum areas such as material investigations and fair testing in Science. They should be encouraged to draw on their own experience and that of those around them, and combine this with information from other easily available sources. For example, when considering making bricks to build his spider's home, Graham looked in the playground to find out about their size and shape and used this information when modelling bricks from cardboard. As Lisa was drawing her initial ideas for a rain-forest shelter, she looked in books to get ideas about what shelters using natural materials might look like. She also talked to others in her class and looked at their models. Yvette wanted to make her roof waterproof and thought back to Science work on materials testing where they had found 'Clingfilm' to be waterproof and see-through and used this to cover her, already painted, roof.

Fig. 5.14 *Yvette used a science investigation of materials to find out how to make her house water-proof*

At Key Stage 2, children should be encouraged to imagine, early on in the activity, what problems they may encounter and what information they may need. Considering *user* and *making issues* (see pages 66–9), should help focus this. The gathering of relevant information can usefully be encouraged by investigating existing products or situations; by making good use of other pupils and/or adults through asking pertinent questions; and by using books and other information sources to support their investigations. For example, a Year 5 class, designing information leaflets for a local leisure site, investigated and discussed a range of existing information leaflets, from many sources, to identify good and bad features. In order to make shoes fit for royalty, Lee measured his own shoe and used this to make a template for developing his design. Combining information from different sources encourages children to be more discerning. For example, Gillian and Gemma made constant use of a history book when working out how to make the masts, rigging and sails for their galleon, but their investigations became more insightful when they combined it with hands-on trial and error of the techniques they read about.

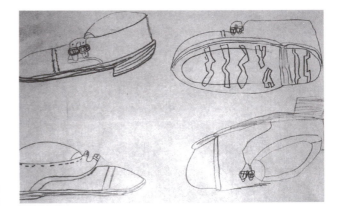

Fig. 5.15 *Close observation drawing of shoes helped the children find out how they are made*

At Key Stage 3, even where tasks are tightly defined and there is a much increased teacher input, it is still important for pupils to consider users, to gather information about their tastes, preferences and needs and to use these in their designing, as Ritesh did by consulting food tables to establish that the energy value of his food product met his client's need. There should also be a clear link between new skills introduced by the teacher and the investigation of their application by the pupil. Noreen (designing and making a textiles board game) was shown how to use a sewing machine and then given time to practise and experiment with the stitching patterns available. Based on this investigation she made decisions about how to use different techniques within her design.

Pupils need to increase their repertoire of techniques of investigating and learn to apply these in a responsive way – gathering information that is genuinely needed. Noreen clarified what was needed through a spider diagram of 'What might a pre-school child be learning?' and a list of 'What research we could do – library, talk to pre-school children, questionnaire'. In order for pupils to tackle such research effectively, both user and manufacturing needs should be presented in a tangible form.

INGREDIENTS	NUTRIENTS		Ingredients	Amount	kcals
3 oz (75g) Plain flour	Carbonhydrates		Plain flour	75g (3oz)	340 kcal
pinch of salt	sodium, potassium, chloride		Salt	Pinch	
50g (2 oz) margerine	fat, Vitamin D, Vitamin A		Margerine	50g (2oz)	730 kcal
25g (1oz) custer sugar	Carbonhydrates		Custer Sugar	25g (1oz)	390 kcal
25g (1oz) Nuts (Hazel)	protein fat		Nuts	25g (1oz)	530 kcal
50g (2oz) Racins	carbonhydrates, vitamins		Sultanas	25g (1oz)	250 kcal

Fig. 5.16
Ritesh used food tables to find out the energy value of his snack bars

At Key Stage 4, it is critical that investigations continue to be directly related to informing decisions about the user and manufacturing issues. Pupils should be able to identify for themselves the aspects of their designing that requires investigating, and then, building on the range of strategies introduced in earlier key stages, carry out research both in depth and with rigour. Giles, for example, when designing and making a kite, identified a need to find out about ways of adding colour to fabric. He investigated techniques through books and instruction sheets and then did extensive tests applying fabric crayons, paints and dyes in a variety of ways to see which provided the best results. In a different project, Kevin realised strength and flexibility were critical aspects of his radio-controlled car chassis and so made up several and tested them by loading them to destruction and recording and comparing the results. In developing ideas for his waterproof 'Walkman' case, Sean investigated a range of user issues, for example to establish the most ergonomic solution to adjusting the controls *through* the cover, and the most effective way of carrying it when engaged in sports activities such as cycling.

Fig. 5.17 *Giles finding out what effects can be created by experimenting with fabric paints*

PLANNING

What is 'planning'?

Planning is any activity which involves working something out in advance of doing it. It occurs at any point in a project, e.g. pupils might plan an investigation, a making process, or a field test of a finished product.

Quality planning includes:

- discussing work as an ongoing part of the process;

- considering time as a resource – and planning its use as a whole and within each phase of the project;

- working out the manufacturing issues before starting construction (e.g. types and amounts of materials, methods of shaping, joining, finishing, etc.);

- careful marking out of materials to ensure accuracy and avoid waste;

- having a view of what the final outcome will be like and how it might be achieved.

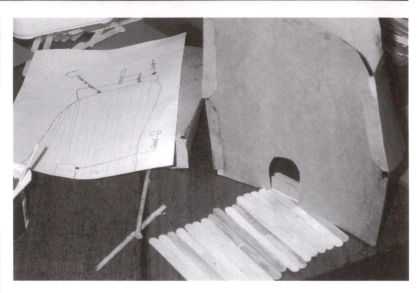

Fig. 5.18 *Key Stage 1 Planning: Lisa and Ashley had their own drawing close by to refer to at all times*

Fig. 5.19 *Key Stage 2 Planning: Lee and Natasha's advance plans for making their shoes*

We are going to make the slipper out of felt. We are going to make the king in stitches. It will be straight stitch. We are going to put se____s round the slipper. Then ___ __e going to put fur inside ___ _ut felt at the botto__ _tasha and Lee Hayward

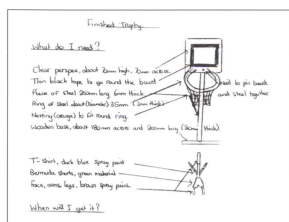

Finished Trophy.

What do I need?

Clear perspex, about 7mm high, 70mm across.
Thin black tape to go round the board.
Piece of steel 250mm long 6mm thick.
Ring of steel about (Diameter) 35mm (3mm thick).
Netting (orange) to fit round ring.
Wooden base, about 180mm across and 200mm long (30mm thick).

T-shirt, dark blue spray paint
Bermuda shorts, green material
Face, arms, legs, brown spray paint.

When will I get it?

green material (for shorts) I have at home, and can be brought in when needed.
Spray paint can be bought on Saterday 11th December, and will be done at home.
Netting - I can get at home, and can be brought in when needed.
The ones missed out are provided at school.

How will I do it?

I'll cut my piece of steel using a hacksaw. Then I will drill a hole near the top. I'll cut my clear perspex to size and drill another hole near the bottom. Then I can put a nail through and screw it up with a bolt.

For the ring I'll use the forge to get it round. Then attach, with glue, the net. Then I'll drill a hole through the ring, make another hole in the long piece of steel and screw in.

Lastly I'll drill a hole in the wood and stick the basket ball net in.   E9

Fig. 5.20 Key Stage 3 Planning: Thomas's plans for making his basketball trophy

Fig. 5.22 Key Stage 4 Planning: Using the blackboard to help pupils keep sight of overall timescales in an 8 week project

Fig. 5.21 Key Stage 4 Planning: Stephanie planning and preparing a mask for her fabric painting

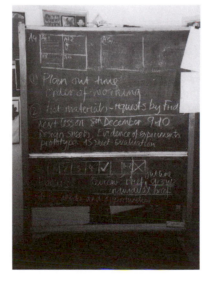

Progressing in planning is supported by:

- encouraging pupils to think things through, take responsibility, and make their own decisions about ways to proceed throughout the activity;

- providing opportunities for pupils to discuss their projects, formally and informally, with each other as they proceed;

- helping pupils to set down their intentions early on in the activity, either through drawing or writing;

- making pupils aware of the time available, and of the time required for particular processes;

- helping pupils to develop strategies for allocating time to the whole activity and its component parts;

- setting intermediate deadlines so that, within a long project, timescales are shorter, more immediate, and therefore more manageable;

- developing an appreciation that a design folder is a working document, and that planning is integral, ongoing and important, not a retrospective presentation exercise.

PLANNING
KEY ISSUES FOR PROGRESSION

With very young children, a plan usually means a drawing of what they want to make, and even though ideas often flow freely and divergently, the step-by-step progress towards the outcome is still generally governed by this initial 'plan' drawing. This does not mean that children do not think ahead: for instance on the day she was to start her spider's home, Emma arrived at school with the shoe box she wanted to start from. They often hold a concept of what 'planning' means – within minutes of starting the shelter project, John declared, 'We need to do a plan of our idea and all the things we need . . . with arrows. It's so complicated!'

This natural willingness to take responsibility for what to do next is a starting point for teachers trying to develop other ways of tackling planning issues. Strategies such as talking about their work, labelling drawings, listing what needs to be done and using further drawings to help visualise what is required, can all help children to think ahead. In this way, planning also becomes an integral part of modelling and evaluating. This was done when Graham wanted to work out how to make a ladder for his spider. He tried drawing a ladder in the air and was encouraged to draw a 'plan'. In doing this he worked out what ladders are like, and then happily set about making one.

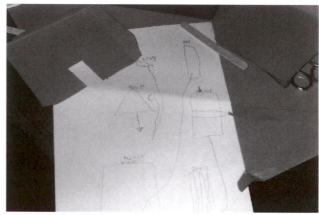

Fig. 5.23 *Planning: David and John's plan 'with arrows' for making their shelter*

The main emphasis in Key Stage 2 is still the 'plan' drawing. But the almost continual discussion (in pairs or group work) of what is required, helps to push the activity forward. However, consideration starts to be given to more complex factors such as the availability and appropriateness of resources, the most effective sequence of activities, and the overall use of time. These can be helped by the use of diaries, logs and pro-forma planning sheets. Leanne used a sheet headed, 'What I am going to do today; what I did today; am I pleased with my work?' to help her decide how much of her roundabout she could assemble in the lesson.

Another activity which helps pupils begin to consider longer term planning is writing 'To do' lists. Sabrina and her partner began the design of their museum exhibit by listing everything that they had to make and/or collect, and only when this was complete did they begin to draw their 'plan'. Importantly, both the list and the drawing were used later to help review progress by allowing the girls to 'see what we've got and what we need'. The teacher's role was largely to monitor discussion, to focus it where necessary, and to suggest strategies to support the pupils' decision-making about time and resources.

Fig. 5.24 *Planning: Annie consults her plan to help her choose the right colour sequins for her shoes*

Pupils in Key Stage 3 devote less time to planning than in any of the other key stages. The teacher-directed nature of many of the tasks means that much of the responsibility for planning is removed from the pupil. In Years 7 and 8 planning tends to be short term and aimed at achieving the immediate task in hand – Paul's planning for his box consists dominantly of marking out the materials that need to be cut.

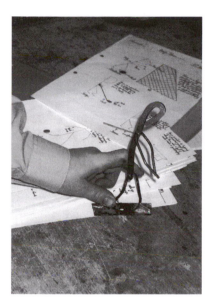

Although the planning of time is usually controlled by the teacher, it is possible to build on pupils' primary experiences of planning, by discussing the targets to be managed during the lesson. This encourages pupils to take responsibility for achieving them. If pupils can be encouraged to utilise their increased knowledge of materials and techniques when planning the use of limited time, then this should help develop skills needed for managing longer, independent projects in Key Stage 4. It is commonplace in Key Stage 3 for project checklists (which detail the aspects to be considered or recorded in folders) to be given out at the start of an activity. These do help pupils to develop a view of the whole activity, but they also tend to encourage retrospective working rather than forward-looking planning.

Fig. 5.25 *Planning: Ben using his working drawings to make sure that his 'golfer's' arms are positioned correctly*

With longer and more pupil-directed activities, the opportunities for planning reappear and the need for it becomes self evident. The teacher's main concern (and usually a major difficulty) is getting pupils to accept responsibility for managing the project and for completing it by the given deadline. Dated time planners, time lines drawn on the blackboard, and 'days remaining' discussed at the start of each lesson, are all examples of strategies used to try to promote an understanding of the project schedule as a whole. However, even with these devices, pupils still find it difficult to allocate their time effectively. Stephanie, when reviewing her work realised that she had 'spent too much time doing the different designs . . . I need to speed up doing my design sheets'. Regular reviews, formal and informal, may be a more effective way of getting pupils to focus on what has been done and what still needs to be achieved. 'Slip charts' can be helpful here, planning (on one side) what *needs to be done* in each session to complete the project and (on the other) reflecting on what was *actually done*. This identifies the 'slippage'.

Pupils appear more comfortable dealing with the shorter-term planning for making, probably because that is what planning typically meant for them at Key Stage 3. Working drawings, cutting lists and templates are pro-

duced prior to manufacture. Sarah-Jane's folder included sketched, full-size, working drawings which showed the shape and relationship of the various components making up the face of her clock. These drawings were then used to produce the cardboard templates from which pieces of acrylic were marked out. Oliver carefully marked out a number of components, using marking fluid, scribing blocks and a surface table, before machining them on the lathe and milling machine.

Fig. 5.26 *Planning: Lisa accurately marking out the position to drill a hole for her clock hands*

MODELLING AND MAKING

What is modelling and making?

Modelling involves the *manifestation* or expression of ideas in order to *develop* ideas. Pupils use two-dimensional and three-dimensional modelling techniques to generate, explore, develop, modify and detail their ideas in the form of discussions, drawings, models, mock-ups and prototypes. 'Making' of the final prototype does not necessarily involve the developmental function of modelling, but is rather the final expression of the design solution.

Fig. 5.27 Key Stage 1 Modelling and making: Using construction kits to model ideas for houses

Quality modelling and making include:

Fig. 5.28 Key Stage 2 Modelling and making: Sabrina working through materials to make a functioning model of a Venetian blind

- pupils talking, discussing (with the teacher, and/or each other in pairs and groups), writing, drawing, and making their ideas;

- hands-on experimenting with materials;

- developing and detailing ideas to the stage where they will work and can be made;

- having a 'big picture' overview of their idea from the start, and then being able to identify and tackle it piece by piece;

- working to clear criteria;

- combining models with drawings and notes to communicate ideas effectively;

MODELLING AND MAKING

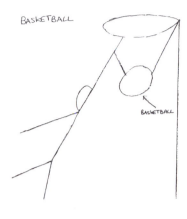 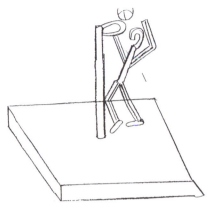

- selecting and using 2D and 3D approaches as appropriate;

- selecting and using appropriate materials and production techniques.

Fig. 5.29 Key Stage 3 Modelling and making: Thomas modelling his ideas for his trophy through drawing

Fig. 5.30 Key Stage 4 Modelling and making: Clive modelling an idea for his wheelchair jack

Progression in modelling and making is supported by the teacher:

- having a wide range of materials, including construction kits, with which pupils can experiment ;

- getting pupils constantly to think and talk about their ideas in relation to agreed criteria for the activity;

- teaching pupils a range of appropriate drawing, modelling and making techniques;

- providing opportunities for a variety of modelling and making techniques to be used within projects.

Fig. 5.31 Key Stage 4 Modelling and making: Clive making his wheelchair jack

MODELLING AND MAKING
KEY ISSUES FOR PROGRESSION

Pupils in this key stage spend a great deal of time modelling – possibly because for this age group modelling is synonymous with making. Pupils have the freedom to produce ideas rapidly, inspired by the materials available (especially reclaimed materials), their own experiences and each other's work. When making homes for spiders the pupils worked largely responsively, referring to a plan initially but then ideas were generated and developed organically from the materials. Yvette started by adding windows to her shoe box and this led to her making a door and then a porch 'just like gran's', with the details being worked out by trial and error and from past experience. In making desert island shelters, pupils worked from 2D drawings to 3D products by hands-on experimenting with materials. John and David also drew on their previous experience of nets and gluing flaps from an earlier project on hot-air balloons.

Pupils also need to be taught how to establish criteria against which their work can be referenced. A discussion between Lisa and Ashley, aided by effective questioning from the teacher, identifies that access to their shelter might be a problem, and that a gate in the surrounding fence could be a solution. As they move from verbal to practical modelling, conceptual understanding about the function and construction of hinges is a serendipitous outcome.

Fig. 5.32 *Yvette's spider's home complete with porch*

As problems become more complex and require more sophisticated solutions there is a need to develop pupils conceptual understanding about materials and to extend the range of communication techniques such as discussing, drawing and making. Gemma was determined to improve the working of the pulley system on her mast and sail assembly, and kept returning to it as her understanding from the pictures in the history book grew. Lee developed his ideas for a 'shoe for royalty' by using his own shoe as a guide when making a paper model from his initial drawing. His consideration of possible colours just before he was about to embroider the shoe, indicates that ideas are still being generated late on in the activity as the immediate making need prompts other thoughts.

A similar approach was demonstrated by Sabrina in her model house. Early on she drew a front and plan view and discussed this with her partner, she then made a card model with the help of the teacher and experimented with ways of joining the material. At a later stage she also modelled a Venetian blind by drawing on her own experience of what such a blind looked like, and through trial and error methods of construction. During the design and manufacture of her fairground ride, Leanne worked from her drawings and experimented with weights and pulleys to see how such a system could be used to provide rotary motion.

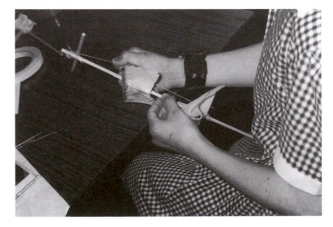

Fig. 5.33 *Gemma working out the pulley system for her sails*

MODELLING AND MAKING

There is a complete change of approach at Key Stage 3 with much less time devoted to developmental modelling (of any kind) and much more focus on making through the instruction and practice of techniques. The approach to modelling is often more formalised ('produce 4 design ideas on this sheet'), with ideas being collected from research material and with the emphasis on drawing as the main modelling tool. Pupils are then typically expected to 'select their best/favourite idea for making'. Thomas used pictures of basketball

players, skiers and other sporting figures to generate and develop the ideas for his trophy, paying particular attention to the posture of the figures and transferring this information to a series of initial sketches, and eventually into well-detailed and annotated drawings.

Using mock-ups and models as a means *for developing ideas* is far less common in Key Stage 3 than in Key Stage 2, but where it happens it can be very effective since the range of materials and techniques is so much expanded. Barry was stimulated to produce ideas by the sudden realisation of the potential of enamelling following a demonstration of the process by the teacher. He sketched his ideas for a brooch, and also experimented with the process to develop them further to see what effects could be created – a combination of working out ideas in advance and developing them through the manipulation of the materials themselves. His excitement with this development process was very evident in his transformed level of motivation for the task. This approach combines the best of the 'playful' qualities of the designer (a really important element of capability) with the rigour that is required of specialist material use.

Fig. 5.34 *Ben making his golfing trophy*

The approach retains much of the formality of Key Stage 3, but becomes more responsive as a result of the individualised nature of the projects. Ideas are developed and modelled in response to the evaluation of real needs and problems as they arise. The initial stages of the process are almost exclusively conducted through drawing and this is then followed by what appears to be a fairly common pattern – sketch, 3D model, detailed working/engineering drawing, manufacture, modify where appropriate. Oliver's telescopic sight mount for an air rifle was first sketched in overall terms and then gradually detailed as the specific requirements of the size of the fixing area and the methods of attachment were established. Once the overall form had been decided, he made a balsa-wood model and tested this on the gun; modifications and developments

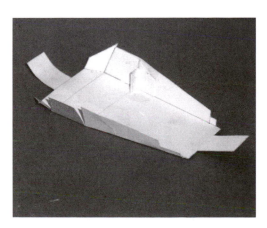

were then recorded in a sketch pad, prior to them being presented in the design folder. Engineering drawings were then produced and used in the marking out and construction of the final artefact. Throughout the whole process individual discussions with the teacher, including a 30-minute review session half-way through the activity, helped Oliver clarify his ideas and push them forward.

Kevin adopted a very similar approach to the development of a chassis for a radio-controlled car. Initial ideas were drawn, then a series of card models were produced and evaluated before the final selection was made and drawn in more detail. The final drawing was very much a 'working drawing' – actually being used on the bench to check measurements and shape.

Fig. 5.35 *Kevin modelled his car chassis in cardboard first*

DESIGN ISSUES
FOR THE 'USER' AND FOR 'MAKING'

What are design issues?

By *user issues* we mean those design issues which relate to the way in which the final designed out-come interacts with users. A child designing and making an educational board game from fabric may be concerned with user issues such as how the game will be played, what will be learned, how it (and the playing pieces) could be stored and how it is kept clean.

By *making issues,* we refer to the design issues concerned with realis-ing the outcome being designed, such as those to do with techniques, tools and materials. The making issues in the board game might involve, for example, what technique to use to attach a pocket for the counters; how to make the fabric stiff enough to stay flat in use but remain flexible enough to be rolled up for storage, and so on.

Design issues arise constantly throughout a project.

Quality consideration of design issues includes:

- considering the user needs and impact of the designing at *each stage* of decision-making;

- demonstrating ability to *compromise and optimise* to achieve an effective outcome;

- children using knowledge of making *as a springboard* for design ideas;

- children using knowledge of making *to evaluate* and support

Fig. 5.36 *Key Stage 1 Design issues: Lisa and Ashley added a fence, gate and booby trap to their shelter to keep the pirates out*

Fig. 5.37 *Key Stage 2 Design issues: Alex and Katie considered both ergonomic and aesthetic issues when designing bike safety wear*

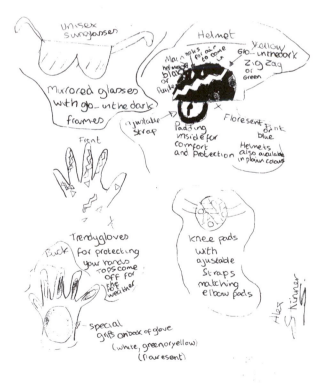

Fig. 5.38 Key Stage 3 Design issues: Barry gained first hand experience of the making issues involved in using enamels

Fig. 5.39 Key Stage 4 Design issues: Sean's early ideas for a waterproof Walkman case show an integration of user and making issues

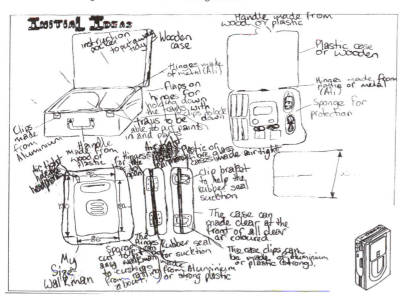

decision-making as they develop their design ideas;

- children *choosing and using materials and tools appropriately* to produce good quality outcomes that work.

Progression in considering design issues is supported by:

- setting tasks in contexts that have a *balance* of user and making issues;

- assisting the learner to engage with the context and take *ownership of and responsibility for* the task;

- helping the learner to *establish clear design criteria* through investigation and discussion;

- helping the learner to keep sight of, and address, the user needs *throughout the project*, including when making decisions about making;

- encouraging learners to use the most *efficacious technique*, material or tool rather than the easiest or most obvious;

- ensuring that learners are introduced to new techniques at the time they are called for;

- helping learners to develop new techniques through 'hands on' experience;

- supporting children to develop an awareness of the fullest range of issues that are both manageable and challenging for them at their stage of development.

DESIGN ISSUES
KEY ISSUES FOR PROGRESSION

At Key Stage 1 children are easily immersed in design contexts, often through fantasy or role play. In the UTA project (Chapter 1), Key Stage 1 children handled user issues more than children at any other key stage. They easily empathise with the 'users' of their designs. This should be built on to encourage them to establish clear criteria through discussion with the teacher and their peers and to keep sight of these while designing and making. Children designing and making the spider's home wanted to keep the spider dry, so they made a sloping roof; they wanted to keep the spider warm, so they made a chimney and fire; and they did not want the home to blow away in the wind, so they put stones in to weight it down.

Making issues can be drawn directly from design ideas, for example considering ways of cutting and joining materials to create the structures envisaged and ways of adding decorative and utilitarian details. They should be introduced to, and encouraged to, consider a variety of techniques, making decisions based on what is best for the job, as Lisa did when choosing between 'Sellotape', PVA and PVA plus paper tabs for joining lolly sticks to make a fence. She chose PVA plus paper tabs because it was stronger and allowed the fence to go round corners.

Fig. 5.40 *Graham works to clear criteria established by the class and teacher for a spider's home that is warm, dry and won't blow away*

At Key Stage 2, Design and Technology tasks should continue to present a broad range of user and making issues. Where it is linked to other curriculum areas (e.g. history) the opportunity to design (rather than just to replicate) should be encouraged. For example, children designing and making 'shoes fit for royalty' in a topic on Tudors and Stuarts, identified issues such as luxury and status, creating footwear that was highly decorated with 'jewels' and embroidery. They also considered more utilitarian issues, wanting the shoes to fit and be easy to put on.

Encouraging children to consider a wide range of user issues, and designing for 'real-life' situations helps to raise the need for *compromising and optimising*. Children designing information leaflets for a new local leisure facility, and safetywear for bike riding, handled a range of consumer issues and had to balance the conflicting demands of fashion against safety.

When dealing with making issues, children should develop a greater concern for the *accuracy* demanded by their more sophisticated ideas and more complex designs. A wider range of techniques will be called for, which can be developed through a judicious mixture of problem-solving and teaching. This was the case with children designing powered vehicles. Sean was taught to make 'Jinks' corners for his vehicle chasis and developed a steering system by problem-solving and pertinent prompting from the teacher.

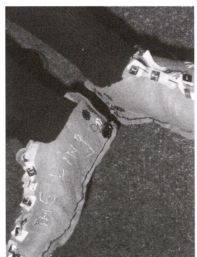

Fig. 5.41 *Shoes beautiful enough for royalty, but comfortable and easy to put on as well*

Whilst the principal Key Stage 3 agenda is strongly influenced by introducing new knowledge and skills, the user and making issues should still be focused. Important user issues identified in the early stages of a project (for example, through brainstorms) need to be kept in sight throughout the project. It is all too easy for pupils to forget them as they get immersed in the making techniques. Whilst Barry was enthusiastically

modelling and developing ideas for the enamel brooch, he completely lost sight of the fact that it had to be worn on clothes. Safety factors (sharp points and edges) were therefore not seen as problematic. On the other hand, Thomas, designing and making an 'action sports trophy', decided that to look good the trophy should really show the action. Decisions on the type of sport and the figure's posture related directly to this issue. Applying newly introduced metal working skills he made a basketball trophy showing a player 'slam dunking' the ball.

Years 8 and 9 should provide increasing opportunities for pupils to use newly acquired knowledge of making as a springboard for ideas. Pupils should be encouraged to handle making issues alongside user issues to improve their skills in optimising, as Hien-Huy did when considering the best way to make his snack bar break easily into bite-size pieces. Knowledge was used to derive a range of ideas, which were then tested against making issues (such as how the bar would release from the mould) and user issues (such as how to make it break without lots of messy crumbs).

Fig. 5.42 *Thomas's basketball trophy showing real 'action' as the ball is slam dunked*

Year 10 marks a transition from the relatively closed briefs of Key Stage 3 to the much more open, personally constructed briefs in Year 11 GCSE project work. Pupils need to develop the confidence to take responsibility for their tasks, identifying and addressing design issues as independently as possible. In a mini–project, Giles designed and made a kite intended to advertise a fair. To capture the traditional feel of fairgrounds, and to make the kite's message visible when flying, he researched traditional Chinese kites. He found ideas for construction and explored ways of creating a bright, distinct surface design. Reviewing his finished work, he identified strengths and weaknesses in his own dealing with user and making issues, and set personal targets for his next project.

Year 11 projects are frequently challenging, placing great demands on pupils. Maintaining a grip on all the design issues is critical, and grappling with them requires confidence, perseverance and a developed skill in

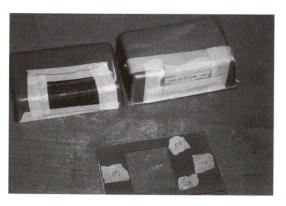

compromising and optimising. Considering ways of attaching his waterproof 'Walkman' case to a waist-belt, Sean weighed ideas for security, strength and protection against comfort, ease of attachment and flexibility. Having resolved this he moved on to deal with keeping the controls dry but accessible, finally creating a simple rubber membrane that fixed easily to the case and allowed access to the controls. Whilst Sean took the lead in this, he used his teacher as a 'sounding board' and support.

Fig. 5.43 *Sean worked on a touch sensitive but waterproof membrane to cover the switches on a Walkman*

EVALUATING

What is evaluating?

Evaluating is an activity which allows the pupil to make a judgement or decision about aspects of the design as it develops, or to reflect on the strengths and weaknesses of the design once it has been completed. It occurs constantly throughout projects.

Fig. 5.44 *Key Stage 1 Evaluating: Yvette evaluating the effectiveness of blue acetate for the windows in her porch*

Quality evaluating includes:

- identifying clear criteria for judgement;

- pupils discussing with each other ways of overcoming problems;

- pupils commenting on their own work and the work of others as it goes along;

Fig. 5.45 *Key Stage 2 Evaluating: A Year 5 group evaluating their sailing ship*

- constantly considering alternative solutions and ways of working;

- using evaluation sheets to review progress and comment on the overall success of the product;

- being disciplined in the collection of evidence, e.g. with question-naires and opinion surveys;

- summative evaluations of the product success against the design criteria;

- basing decisions on a wide range of available evidence, including prior knowledge, testing, trial and error, etc.;

DESIGN IDEAS Pg 4

[handwritten design ideas and evaluation notes]

Fig. 5.46 *Key Stage 3 Evaluating: Seema designed, made and evaluated three different snack bars before her developing her final idea*

Fig. 5.47 *Key Stage 4 Evaluating: Stephanie used the attainment targets to provide a framework for evaluating her beach bag project*

[handwritten evaluation notes]

- using different kinds of evidence – qualitative and quantitative – distinguishing fact from opinion etc.;

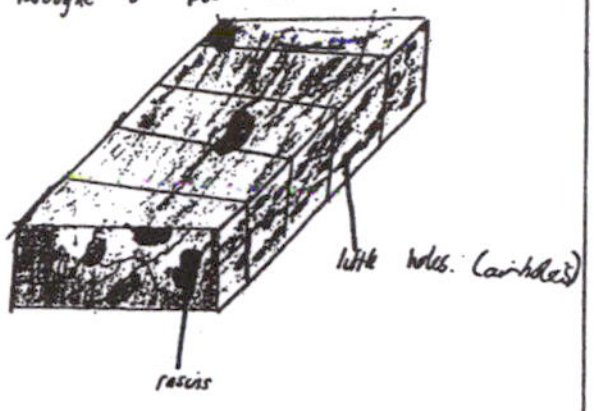

- resolving conflicting demands to produce an optimum solution.

Progression in evaluating is supported by the teacher:

- creating specific opportunities for pupils to discuss work one-to-one with them, within groups, and as whole-class activities;

- prompting pupils to develop design criteria at the start of an activity;

- prompting pupils constantly to compare their products against these criteria;

- encouraging children to keep these criteria under critical review so that they can be modified as the project develops;

- encouraging pupils to consider the user of the product, and ensuring that this is a key factor in its evaluation;

- helping pupils to develop techniques for gathering reliable data for evaluation;

- guiding project selection to ensure an appropriate level of challenge;

- allowing pupils to take responsibility for as much of the project as possible;

- teaching pupils a range of strategies to evaluate their work.

EVALUATING
KEY ISSUES FOR PROGRESSION

At Key Stage 1, evaluation occurs naturally and consistently throughout the life of projects in response to the development of ideas and solutions, often prompted by the teacher, but usually in the form of almost constant discussion and comment on each other's work. This commentary is nearly always given and accepted in a supportive manner – when designing and making homes for spiders, Emma told Yvette that the paint she was just about to use was likely to run, and told Henry that pink was not a suitable colour for boys!

Evaluation results from 'hands-on' experience of using materials and having to modify ideas and plans by trial and error. Lisa and Ashley realised that the card they had cut for the base of their model tree was too small and too stiff, and therefore they went on to try tissue paper. John wanted to lift his desert island shelter above the ground on stilts. His first attempt was too wobbly, and he had to reconsider the materials and the method of construction.

Pupils respond to problems as they arise, but see less need to make summative evaluations of their finished product. To begin to develop this aspect, teachers need to help pupils develop design criteria at the start of the activity, and then allow time for discussion at the end which focuses on the degree to which these criteria have been intentionally modified or met as originally stated.

Fig. 5.48 *Extract from a taped conversation with Lisa, Ashley, David and John evaluating their shelter*

Lisa:	I think it's good [of David and John's shelter] because they put, umm, they managed to make the plank and the hook. I think [ours] is not very good because I didn't get to put the same kind of yellow . . . I would like it if we had the same kind of green, but in the end it turned out grey.
John:	I like the way Lisa's drawn the grains of sand . . . and I like the logs so the pirates will slip and slide on them.
Lisa:	I think it would be good if they [John and David] had a door that opens and closes.
David:	Um, if you done a door that opens and closes and painted it brown, if the pirates got up the log, they'd think it was just a wall, and they'd just get locked out.

Key Stage 2 pupils have considerable ownership of tasks and autonomy of decision-making in activities and therefore the evaluation which occurs is a natural and critical response to their developing ideas. For example, Sean changed his vehicle design when he realised that the axles were not running as smoothly as they should, and he looked again at how they were attached to the chassis; Gemma devised a simple hinge system to allow a section of the railings on the galleon to be removed when she realised that the sailors and barrels of provisions would be unable to get on board; and Lee tested the size of the shoes he had made by trying them on.

More formal evaluation strategies also begin to be taught, with proforma planning and evaluation sheets and structured group and whole class evaluation sessions. All of which can help pupils check the success of their work when compared to the design criteria.

Pupils are also becoming more aware of the design and make process and how its effectiveness can vary. Gillian was able to comment at the end of the activity that although they had worked in pairs to design and make parts of the galleon, there had been little if any communication between the pairs, and therefore there were parts of the model which did not fit together very well.

Fig. 5.49 *Ongoing evaluation to check that the shoes will fit*

At Key Stage 3, since there is less autonomy in decision-making (and often less ownership of the tasks), evaluation is too often seen as a necessary burden at the end of the project rather than as a critical developmental force throughout a project. If the level of decision-making is superficial (because the task does not allow room for it), then there is relatively little to be lost through inadequate evaluation. This was why several of the enamel brooches were able to escape critical scrutiny and end up with dangerous points and edges. But if the pupil has to make major decisions about the appearance, function, materials and construction, then inappropriate choices have immediate, and very tangible, consequences for the overall success of his or her solutions.

Overall Key Stage 3 has the least evaluation of all the key stages, with Years 7 and 8 having less than Year 9 where the activities are becoming somewhat less directed. Most evaluation occurs summatively, and the ongoing evaluation that does exist, typically relates to the processes and quality of manufacturing (Thomas repeatedly checks his progress in bending the metal for his 'action sports trophy', by comparing it with his drawing.)

Formal written evaluations at the completion of the activity tend to be treated uncritically and are often regarded as of little relevance, though some pupils in the UTA study were able to describe the importance of this part of the activity in helping their thinking, decision-making and designing; as was the case with Richard (Fig. 5.50).

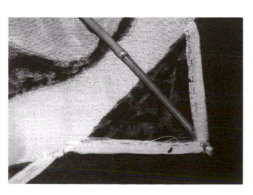

Fig. 5.50 *Richard's final evaluation of his basketball trophy*

Evaluation at Key Stage 4 almost matches the amounts conducted in Key Stage 2, but, unlike this earlier key stage, where it is spread relatively evenly throughout the activity, the vast majority is conducted towards the end. There is evidence in Year 10 of a growing awareness of the importance of ongoing evaluation, although much of it still tends to be of a trial-and-error nature. Giles, for example, tested the relative effectiveness of fabric crayons and paints for his kite design, and also tried a number of ways of stitching the corner pockets. Self-assessment sheets were frequently used to encourage evaluation of the process and the product.

In Year 11, pupil-selected projects allow the pupils to take full ownership and control of projects, and therefore they see the critical importance of making the correct decisions. Pupils are often in the position of making decisions for which there is no known, certain answer. Sean needed to make a flexible cover for the switches of his 'Walkman', which was waterproof, but which would also allow the switches to be operated. This involved him in a complex range of decisions involving the size of the switch cover, the type of material and the methods of fixing it in place. He only achieved this by evaluating alternatives and trying to optimise his decision.

Through the misconceptions of many GCSE examination formats, the vast majority of marks for evaluation in design folders tend to be awarded for summative rather than ongoing evaluation, and the final page of the folder almost becomes a ritual to be completed rather that an important aspect of the whole process. Nevertheless, it can be enormously strengthened by drawing evidence from external sources with expertise in the area, e.g. it would have been interesting to read an evaluation from Sony of Sean's 'Walkman' adaptation.

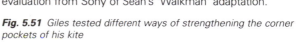

Fig. 5.51 *Giles tested different ways of strengthening the corner pockets of his kite*

EXTENDING KNOWLEDGE AND SKILL

What do we mean by extending knowledge and skill?

The knowledge and skill we are specifically concerned with in Design and Technology is that which can be used to make things work:

- to work technically – the materials, components and techniques applied and the systems created;

- to work for people (ergonomically) – creating outcomes that fit people – both physically and in their day-to-day lives;

- to work aesthetically – the form outcomes take, their quality and the impact this has on our senses and our sensibilities.

In the APU project (see Chapter 1) we constructed a simple way of judging the level at which knowledge and skills were being used:

- At a *black box low level* of operation, e.g. where a pupil says something will work 'by clockwork' or that the material needed is 'grippy stuff', or that something should be 'joined', with no further details to qualify the proposal. Things work as if by magic.

- At a *street level* of operation, where knowledge is consistent with that held by the person in the street, but not developed by exposure to specific teaching and learning in Design and Technology.

Fig. 5.52 Key Stage 1 Extending knowledge and skill: Emma uses previous knowledge of using tabs to fix the chimney to her model

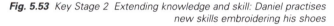

Fig. 5.53 Key Stage 2 Extending knowledge and skill: Daniel practises new skills embroidering his shoes

74

EXTENDING KNOWLEDGE AND SKILL

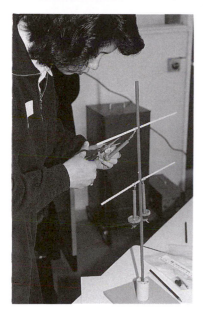

Fig. 5.54 *Key Stage 3 Extending knowledge and skill: following her teachers input, Nasreen experiments with ways of moving joints*

Fig. 5.55 *Key Stage 4 Extending knowledge and skill: Kevin assembles his car chassis, with support from his teacher*

• At a *working knowledge level*, where specialist Design and Technology knowledge or skill is developed and applied with understanding to fit the particular requirements of a proposal.

We will use these terms to refer to development and progression in this area across the key stages.

Quality extension of knowledge and skill includes:

• operating at a level appropriate to the stage in designing – black box may be appropriate when preliminary ideas are flowing in a divergent way, but working knowledge will be necessary when the detail of the designing and making is being considered;

• working from what you know, through what you need to know to the acquisition and application of new knowledge and skill;

• operating through first principles to identify ways of resolving a task;

• using knowledge and skill to make good-quality outcomes, fit for their purpose;

• applying established aspects of knowledge and skill in new settings.

Progression in knowledge and skill is supported by:

• hands-on experimentation and problem-solving, resourced by judicious prompting and questioning from the teacher;

• hands-on activity to reinforce, practice and clarify new knowledge or skill;

• drawing on children's previous experience, transferring knowledge, skill and understanding into new settings;

• children reflecting on what they know and can do and how they can make best use of this;

• encouraging children to take responsibility for both the thinking and doing involved in their task;

• providing pupils with opportunities to take control of their learning;

• pupils constantly being required to think ahead about the areas of knowledge and skill with which they will need to come to terms;

• encouraging children to leave their 'comfort zone' of established knowledge and skills to tackle challenges that force them into new areas.

EXTENDING KNOWLEDGE AND SKILL
KEY ISSUES FOR PROGRESSION

KEY STAGE 1

At Key Stage 1 much development of knowledge, skill and understanding comes through hands-on investigation, which highlights the importance of experimentation and play. In the UTA project, the children we observed were generally operating at 'black box' levels of understanding, although one or two did indicate a more 'street level' understanding of the needs of people. In some instances, children were operating 'pre-black box', for example a group of children developed a concept of 'hinge' through serendipitous experimentation with materials, even though they previously had no knowledge of the word 'hinge'. They discovered it by joining lolly sticks with paper to give them movable sections, incorporating this into a gate for the fence in their shelter. It was important that the teacher then reinforced their discovery by providing the appropriate technical word.

Introducing new knowledge or skill when it is directly needed, allows children to internalise, practise and make sense of it. They can also be encouraged to draw on other experiences, transferring knowledge and skill into new settings. A child who wanted to make a shelter waterproof, thought back to Science experiments and suggested that candle wax or shoe polish could be used. Another child wanted to add a chimney to her shelter and drew on previous skills in joining cylinders to flat shapes using tabs to do this.

Fig. 5.56 *Lisa and Ashley discover a way of making a hinge for the gate in their fence*

KEY STAGE 2

At Key Stage 2, children need to be involved in more complex designing and making that will push the knowledge and skill demands towards working knowledge in the specific areas of their project. This was exemplified by Sean, who in designing and making his powered vehicle, used knowledge to help him decide whether to make his vehicle front or rear-wheel drive, basing his decision on the size of the wheels and the weight and position of his battery pack. Sean was working within a common Key Stage 2 model for development, where problem-solving is blended with teacher inputs when the children get 'stuck'. This is further exemplified by children trying to create a stable structure for their model house. The teacher introduced them to the concept of slotting sheet material together, and they then worked by trial and error to achieve a system that was reasonably stable. The teacher provided extra guidance to help increase the structural stability.

As with Key Stage 1, building on previous understanding and introducing new knowledge and skill at critical moments is important. It is vital that pupils are engaged in designing and making that is just within or just beyond their reach. This challenges them constantly to extend into new understandings in order to achieve success.

Fig. 5.57 *Sean's rear wheel drive car*

EXTENDING KNOWLEDGE AND SKILL

In Year 7, knowledge and skills are introduced in an intensive way, and generally in bite-size chunks of working knowledge. In order to ensure that pupils develop understanding along with knowledge and skill, it is important that new knowledge and skill is introduced at a point where pupils can take responsibility for applying it in their designing and making. This can be illustrated by Barry, who having been introduced to the technique of enamelling, practised this on his original idea for a brooch and then used his new understanding to design and make a range to extend his first idea. It is also important that pupils are involved in Design and Technology tasks that require them to draw on previous understanding, as Hien-Huy did when designing his snack bar, drawing together his understanding of nutrition, shaping material and creating a product pleasing to eat.

Knowledge and skill that is introduced by direct teaching and followed by hands-on experience to explore it can ensure that pupils receive quality information. However, in order for it to get beyond 'inert' (lifeless) knowledge it is important that such teaching is linked to the pupil's 'need to know' for the project, and with the opportunity and requirement for the pupil to take some responsibility for its application in the project.

Fig. 5.58 *Barry's brooch (top right) proved a starting point for enthusiastic experimentation wiith enamelling*

At Key Stage 4 pupils should be given increasing responsibility to manage design projects. This requires them to draw on previous understanding and seek out new knowledge and skill as demanded by the task. This was the case with Stephanie who designed and made a beach bag, drawing on her previous understanding of creating surface decoration on fabrics. In the process she transformed a series of former black box understandings by using instruction sheets and first-hand experimentation and developed some sophisticated working understanding of the behaviour of textiles.

Individual projects in Year 11 present a wealth of opportunities for pupils to consolidate previous learning and develop real depth of understanding through the demands in their tasks. Teachers of this age group are well aware of the potential growth in some pupils at this stage. Development should be promoted across the

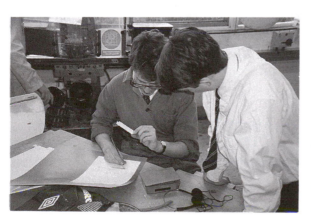

key stage by encouraging pupils to take responsibility for achieving their design intentions throughout all project work, pushing rigorously at getting ideas to work. In facilitating the pupil, the teacher has the tricky job of providing support without creating dependence. We observed some amazing examples of this – teachers providing half answers dripping with clues that enabled the pupils to take them forward independently. This enabled pupils to glory in their personal success when they made it work, unaware of the subtlety of what the teacher had done.

Fig. 5.59 *The teacher acting as facilitator to enable Sean to move forward confidently with his Walkman case*

COMMUNICATION

What is communication?

Communication has two different aspects in Design and Technology:

- communication to *clarify* and develop ideas;

- communication to *record and present* ideas.

These two dimensions form the focus of the work described here at each key stage.

Fig. 5.60 Key Stage 1 Communicating: Sally and Rebecca discuss their evaluation

Quality communication includes:

Fig. 5.61 Key Stage 2 Communicating: Chuks and Kun debate the construction of their model

- the many kinds of drawing, modelling and discussing that supports design thinking and the development of ideas;

- the use of drawing and modelling to enable pupils to transform ideas into different arrangements;

- the exploitation of the different strengths of each form of communication
  - verbal communication for speed;
  - graphic communication for style and relationships between parts;
  - construction drawing for detail and precision and concrete modelling to test if it all fits together;

- presenting ideas and outcomes with clarity and skill.

78

COMMUNICATION

Fig. 5.62 *Key Stage 3 Communicating: Danny consults Faith and Lois on his next step in modelling ideas for play equipment*

Progression in communication is supported by:

- encouraging pupils to engage in dialogues as a natural part of designing, using sketching, discussion and other forms of modelling;

- introducing drawing and modelling skills and formal conventions when they will directly support the designing;

- encouraging pupils to record work in such a way that it supports consolidation and reflection;

- providing pupils with a repertoire of communication 'tools' and encouraging them to use these appropriately.

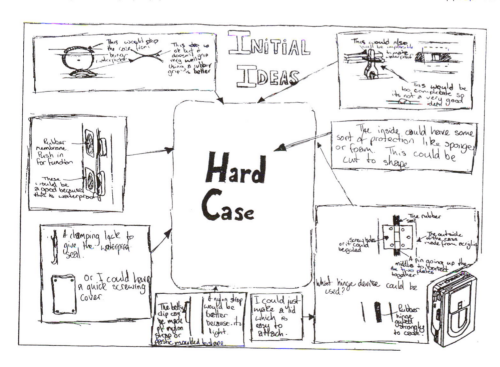

Fig. 5.63 *Key Stage 4 Communicating: Sean exploits sketching techniques to present early ideas for aspects of his Walkman case*

COMMUNICATION
KEY ISSUES FOR PROGRESSION

At Key Stage 1 children can be supported in their use of communication to clarify and develop ideas through discussion (with the teacher and with their peers), through representation of their early ideas as a drawing or 'plan' and through the outcome they produce. Ideas can be dealt with in a very immediate manner and, where children are working on their own tasks in an informal group, ideas can be discussed and evaluated by the children as they are working. An initial plan or drawing of their idea is a valuable way of helping them rep-

resent their thinking. Sometimes this bears a close relationship to the final outcome, even if ideas may have been modified and added along the way. Sometimes the initial drawing holds little explanation to the adult eye, but is kept in sight and referred to regularly by the young child. Encouraging pupils to make a report of their project – showing stages of making and the decisions made in producing the outcome – is a valuable tool for reflection and consolidation as the child looks back over what has been achieved.

Fig. 5.64 *Emma's plan for her spider's house*

At Key Stage 2, planning, drawing and discussing tend to be used interactively as pupils clarify and develop their ideas. This was particularly evident with Lee and Natasha who jointly worked on designing 'shoes fit for royalty'. Although each had their own design sheet, they worked in tandem, constantly discussing ideas and leaning across to each other's sheets to make notes and add detail to the ideas being developed. As children develop more complex ideas, they should be encouraged to use separate models to explore and communicate their thinking. Where children were making a model supermarket for their museum exhibition, they needed to model their ideas for display cabinets to sort out both the proportions and ways of fixing the units to the floor. With the group of children designing public information leaflets, before any drawing took place,

they modelled (by paper-folding) their ideas for folding the leaflets and for laying out the information.

Preparing a project report is valuable, and good use can be made of information technology for the presentation of both text and drawings. Encouraging pupils to use logs, diaries and planning sheets is a useful way to encourage reflection, both through the project and at the end of it.

Fig. 5.65 *Lee and Natasha discussed their plans at every stage*

COMMUNICATION

At Key Stage 3 pupils are typically introduced to more formal methods of communicating, both for the development and the recording of ideas. This can be important in helping them address the increasing complexity in designing and making. This has been found to be most effective when formal skills have been introduced directly in relation to a design development function, for example introducing orthographic drawing when pupils would otherwise be struggling to work out the dimensions of a 3D product. Encouraging pupils to keep a project folder can be useful for consolidation and reflection and can help pupils develop a pride in their work, but pupils must see a genuine need for folder work in terms of the design development if it is not to become a superficial and retrospective recording, masquerading as more immediate reflection.

Fig. 5.66 *Being introduced to formal drawing techniques helped Richard finalise shapes and sizes for his trophy*

At Key Stage 4 project folders take on even greater significance as they increasingly are seen as providing evidence of the pupils' design thinking for assessment purposes. In order that time is spent productively, it is important that all folder work genuinely supports the development of the project, and that pupils use communication techniques appropriately. Such a responsive approach was witnessed in one school where work was carried out as appropriate to the immediate needs of development (through rough sketches, samples of printing, collections of visual resources, etc.) and then mounted onto a set of presentation boards telling the story of the project. This method produced an effective record of the project that the pupils felt had not taken an undue amount of time.

It is important that pupils develop free-hand drawing skills to express, develop and clarify their ideas. They should also value them, and not feel that they need to be redrawn 'neatly' for examination purposes. Sketching in a notebook, folder or sketch book can be a personal monologue (thinking with oneself) and it often forms the centre of a dialogue between teacher and pupil as ideas are explained, explored and developed. As the project moves forward, these quick freehand sketches need to be supplemented with more formal and detailed styles of communication that allow ideas to be examined more rigorously. But sketching, discussing and 3D modelling should continue to be encouraged.

Fig. 5.67 *Stephanie's early ideas mounted for presentation*

Some principles of progression

In this chapter we have attempted, through the use of UTA case study material, to show examples of design and technology capability in each of the four key stages. We have also explored a range of processes, knowledge and skills which are the constituent parts of that capability. It is only by coming to understand the role and interrelationship of these individual elements that it is possible to clarify what represents a quality piece of work. It also then becomes possible to see how teachers can plan a progressive and coherent experience that may help pupils to develop towards capability.

However, it should be apparent that there are some overarching issues which directly affect the performance of pupils in a number of these aspects of capability, and it is these that we shall summarise here.

More seems like less

Progression might seem a straightforward concept – you learn something, then you learn a bit more, then a bit more, and so on. Through this process of accumulation, one's knowledge, skill and understanding grows and we become more competent and capable. One of the sad facts about progression, however, is that the more we learn about something, the more we realise what we do not know. This uncomfortable truth can come to children at a very young age. Penny Munn, in researching children's understanding of reading found that in a nursery of four-year olds, at the start of the year all said they could read. By the end of that year only a small number still maintained they could read. The others had all come to see that there is more to reading than sitting on dad's lap, looking at a book and having a story! (Munn 1995). So a critical component of progression involves the conscious realisation that we cannot do something, or at least that it is more complicated than we first thought. It thereafter requires that we put ourselves in the firing line, operating with uncertainty and taking on new challenges.

Concrete and abstract challenges

We have argued in Chapter 3 that one of the defining characteristics of Design and Technology is that it enables pupils to gain access to complex concepts through concrete means. We argue that this concrete access enhances pupils' learning opportunities, and over the years, many educational writers have made this point for us. From Piaget (1958) (who distinguished 'concrete' and 'formal' operations), to Bruner (1968) (who distinguished between enactive, iconic, and symbolic modes of operation) and Donaldson (1978) (who argued for an 'activity' model of curriculum), we have become increasingly aware of the importance of concrete, first-hand experience for children struggling to make sense of newly introduced concepts. This has recently been taken a stage further by Rogoff (1990), whose arguments about 'situated cognition' suggest that children's learning is not of generalised qualities but is specific to the concrete context within which they learn it. To be able to generalise such learning would therefore be a mark of progression. This is further illuminated by the work of Clare and Rogers (1994) who established the importance of 'metacognitive' reflection in progression. As children are encouraged to think back over their work and turn tacit concrete operations into explicit understandings, they can make them more robust and more transferable to new situations.

All of this underlines the tight and complex interrelationship between concrete activities and pupils' progression in learning and this relationship can be seen at work in these case study materials.

Sequential and organic models of progression

It has frequently been argued that progression is sequential: first learn to do X and then it is possible to do Y. It appears to be common sense, and in Design and Technology this argument has been used to justify the view that children must *first* learn the essential tool skills, and *then* (and only then) can they decide how they might want to use them on projects of their own. More recently, it has been recognised that such 'basic skills' programmes completely ignore the need for children to see (as early as possible) the necessary connection between means and ends. To learn a collection of isolated, decontextualised skills will not help children to recognise when they need to deploy them. Accordingly a different paradigm of progression might be advocated and it would start from a very simple principle. Anything that pupils are expected to be able to do at the age of 16, they need to be getting started from the very beginning. As Bruner puts it;

> Any idea or problem or body of knowledge can be presented in a form simple enough so that any particular learner can understand it in a recognizable form.
>
> (Bruner 1968)

This model of progression suggests that learning amounts to progressively peeling layers of consequence and meaning. A six-year old can understand anything at a six-year-old level, and a 16-year-old can understand the same thing at a much deeper level. This is after all how we manage the questions that our children ask about human reproduction. An answer that we might give to a 6-year-old will only last for a while, and we progressively have to make our answers more complex as their understanding deepens. It is easy to see how such a principle might apply to Design and Technology. If cutting and forming sheet materials is an important concept, it is easy to exemplify through 6-year-old levels of making as well as 16-year-old levels. The materials and the tools might be different but the concepts of cutting and forming can be common. Similarly, if the liquids-into-solids transition is important to understand and experience, chocolate and jelly make excellent media through which to introduce it in an immediate form with very young children.

If we wish children to progress, we need to present them with activities and challenges that enable them progressively to peel these layers of understanding of the technological world. Again this has emerged through the case studies. All the pupils (for example) have a view about how materials behave and how they can be manipu-

lated and joined. As they get older, these same concepts are being explored, but in progressively greater depth.

Balancing risk

Learning in schools presupposes teaching, since teachers have generally more expertise than the children they teach. However, one of the most difficult skills that our student teachers have to learn is how to restrain themselves from telling children how to do things. They have to learn to rein in their natural tendency to give children answers all the time, and rather encourage the children (through questioning and other techniques) to work out a solution for themselves. Any experienced teacher knows that receiving pre-digested truths from the teacher is far less demanding on children (and hence less extending of their intellect) than helping them to struggle to formulate such truths for themselves.

This is just as true in Design and Technology as it is in any other area of the curriculum. However, once again in Design and Technology, the significance of it may be more immediately apparent. The basic *modus operandi* in Design and Technology is to use techniques of 'designing' and/or 'problem-solving' and here it is patently necessary for children to operate at a level of uncertainty. They are in pursuit of a solution that has not been pre-packaged for them, and many different solutions might work. This is a potentially stressful and certainly a risk-taking environment. Teachers could remove this stress and eliminate the risk by using their expertise to give them all the answers, but that would destroy the whole purpose of the exercise. As Dewey puts it

> A difficulty is an indispensable stimulus to thinking, but not all difficulties call out thinking. Sometimes they overwhelm and submerge and discourage ... A large part of the art of instruction lies in making the difficulty of new problems large enough to challenge thought and small enough so that in addition to the confusion naturally attending the novel elements, there are luminous familiar spots from which helpful suggestions may spring.
>
> (Dewey, 1968)

Children's progress relies centrally on the teacher's expertise in manipulating the level of challenge in tasks and in striking a balance that requires pupils to take risks – but only those that are carefully calculated to be manageable. This point has emerged clearly in the case studies, particularly in the contrasted models of teaching and learning in Key Stages 2 and 3. We shall explore this in more detail in Chapter 7.

Discussion (verbal modelling)

The way in which discussion can enhance pupils' performance was initially highlighted by the APU research findings, and we have outlined the issues in Chapter 1. The UTA findings completely confirmed this critical role, but also highlighted that it was only primary pupils who regularly and consistently make use of it. In primary schools the technique is used formally and informally, as pupils discuss with each other:

• what they are doing;
• why;
• how they are going about it;
• what they have completed so far;
• what they need to do in the future;
• the problems that they have encountered;
• the ways in which they have overcome them; and
• the success or otherwise of their work.

Secondary pupils find themselves in a very different position. Partly because of the individualised nature of much project work and partly because of the pressure to teach knowledge and skills in a limited amount of time, it is very rare to see Key Stages 3 and 4 pupils making use of a technique that was at one time second nature to them. Consequently, the majority of secondary pupils are denied access to a vital tool which can help them clarify their thinking and develop both the active and reflective aspects of capability.

It is worth recalling the pupils' response to the questionnaire that was returned by all pupils after they had completed their APU activities. We had used a series of strategies in the activities and we asked pupils to rate them in terms of how helpful

they had found them. The response in relation to the discussion session speaks volumes.

Table 5.1

No help	8.60%
Some help	34.32%
Very helpful	56.21%
No response	.87%

Note: Total sample 1288 pupils
Source: Kimbell *et al.*, 1991, section 8

This was from a group of 15-year olds with almost no recent experience of using the technique. There are few things in life that are certain, but one of them is that the occasional use of paired and/or small group activities, and the use of short but structured discussions to review progress, will enormously help pupils to make progress in their work.

User issues

The consideration of user issues has an enormous impact on a number of aspects of capability. If they are ignored, or only treated in tokenistic fashion, the performance of the pupil is likely to be seriously impaired. When the pupils realise that as well as designing and making X, to do Y, they are also doing this for person(s) Z, the activity takes on a whole new level of meaning. If the outcome is to be successful it now has not only to do Y, but it has also to do it to the satisfaction of Z. Consequently, the user dimension has to be considered at every stage of the process:

• setting the design criteria which the outcome has to fulfil;
• collecting useful and relevant research information;
• evaluating ideas as they are modelled and developed;
• and testing and evaluating the outcome.

Key Stage 3 is an anomaly when it comes to user issues. Here the consideration of the user comes second to the demands of making, and to the development of new skills and knowledge. The

UTA projects suggest that, although there are very few Key Stage 3 activities where user issues are completely ignored, it is equally true that there are very few where user and making issues are treated in a balanced or integrated way. In many cases all it requires is to stress the importance of the user, and not allowing this to be lost as the activity progresses; in others, it may require a re-appraisal of how the activity is presented – or even how it is designed.

Working through materials

In every key stage, without exception, it is possible to see the importance of pupils working 'hands-on' with materials:

- to investigate and model their ideas;
- to evaluate alternatives;
- to establish the suitability of a material for the job;
- to develop an understanding about material properties; and
- to help clarify and consolidate their grip on new knowledge.

The ability to push ideas forward by trying them out in a concrete way is one of the unique features of Design and Technology activity, because pupils are confronted with the consequences of their decisions in a very tangible form. Reflection is enhanced and complemented by action, and it is the teacher's role to try and ensure that the pupil has the necessary opportunities and skills to do both.

Ownership of the task

It is self-evidently a good thing that pupils should take ownership of the tasks on which they are engaged. In Chapter 4 we showed how the structure of the activity and its position within a hierarchy of tasks will affect pupils' ability to do this. The earlier in the activity that pupils can take on this ownership, the more effective will be their overall response, since personal interest and involvement will drive them forward. Encouraging pupils to make their own informed decisions, and to take

responsibility for the planning and organisation of the activity, are all necessary attributes of capability.

This is most difficult to achieve in Key Stage 3, where the teacher is seeking to retain control of the project in order to use it to teach some specific elements of knowledge and skill. This is particularly true at the beginning of this key stage in Year 7, where the increased amount of teacher direction, and the knowledge/skills-driven nature of the tasks seriously limits the opportunities for pupils to demonstrate autonomy and develop ownership of the activity. Once again, this breaks a progression which starts at Key Stage 1 with pupils naturally empathising with the users of their design work, and finishes at Key Stage 4 with pupils simulating 'real' technology in projects which they have chosen, and for which they can see a real purpose. It is a real challenge to Key Stage 3 teachers to find ways of working within these constraints which will provide increased opportunities for pupils to feel that they can exert some influence over the activity. This can be assisted by setting tasks in user-rich contexts and encouraging the pupils to take responsibility for the planning and management of their work within the framework of constraints established by the teacher.

The transition from Key Stage 2 to Key Stage 3

From the UTA project observations, it became clear that in many respects Key Stage 3 represents a discontinuity in the progressive development of design and technological capability. The full implications and reasons for this will be discussed in Chapter 7. However, it needs noting here because of its apparent impact on progression. Because the UTA study was not longitudinal it is not possible to comment on the performance of pupils as they move from the primary to the secondary phase, but the discontinuities can be expected to have a serious impact on pupils' performance.

We are not saying that the nature of Design and Technology activities in Key Stage 2 are better than that in Key Stage 3, or vice versa, just that they are different. And *very* different. But having identified here some characteristic and effective ways of working in these key stages, it may be worth considering how the best practice of each could be incorporated into the other. For example, secondary teachers might benefit from seeing how their primary colleagues organise groups, encourage pupils to discuss their work, integrate user and making issues, and take responsibility for planning and organising their work. Equally, primary teachers might gain from seeing how new skills and knowledge are taught through secondary project work, and the range of strategies employed for generating, developing, recording and evaluating ideas and outcomes.

Above all, there appears to be a need for primary and secondary teachers to meet, talk, observe, review each other's teaching methods, and learn to take the best from each other's approaches. This is obviously not a quick, easy or cheap exercise to organise or implement, but if it were possible, if only on a limited scale, the results would surely be worthwhile. We look in more depth at the issues in transition from primary to secondary in Chapter 7, but before doing this there is a further area of importance to be considered.

Over the preceding pages we have illustrated differences between key stages and at times within key stages, but as yet we have not dealt with differences between individual pupils. And as we know, all pupils are different and respond differently to the challenges we present to them. We are raising here the critical area of differentiation, and it is to this that we must now turn.

SUMMARY

- Through specific case study projects we outline some *overview characteristics* of pupil performance to exemplify what we mean by quality work at each key stage, enabling us to build a picture of progression.
- Through selections of case study materials, we identify a series of facets of performance that we believe to be central to the development of children's capability. We exemplify and discuss:

> investigating
> planning
> modelling and making
> design issues (user and making)
> evaluating
> extending knowledge and skills
> communicating.

In each case, we define and exemplify each of these facets of capability and identify some key indicators of quality. We then show how this quality progresses across the four key stages.
- Finally we examine some broad principles and common issues that emerge as being particularly significant for the development of capability.

CHAPTER 6

All children are different: the challenge of differentiation

Introduction

Differentiation is about being aware of the differences between learners and taking account of these differences in our teaching. It follows that – as we indicated in Chapter 1 – we can never expect to appeal to the 'right' way of teaching anything. Such is the magnificent diversity of humankind, what is right for one learner in any given setting will be wrong for another.

Differentiation is ultimately an 'equal opportunities' issue. All young people should properly expect to have equal access to learning in Design and Technology. The irony of this, however, is that in order to provide equal opportunities, we must provide an unequal teaching and learning diet since a standard diet would be guaranteed to penalise some and favour others. A commitment to equal opportunities in education is ultimately a commitment to an individualised curriculum:

> The paradox of equality in education is that it is only when the educational diet of every child is different from that of every other, that we can really hope that we are near to achieving it.
>
> (Downey and Kelly, 1979)

In reality it is very difficult to manage an individualised curriculum when one is responsible for 30 children at the same time. What we can do, however, is to be aware of the impact of the strategies we are using on the different styles of learning that are represented in our classrooms.

There is a further issue here, for a differentiated curriculum must not only recognise different starting points and different styles of learning, it must equally recognise that we do not want all learners to end up the same. We do not need a nation of clones, all conditioned to respond in predictable ways. In this sense, the curriculum is not about convergence towards an ideal state. This is not an anarchistic declaration that 'anything goes', for we have already outlined in Chapters 2 and 3 what we mean by technological capability and to that extent we would expect all pupils to share certain qualities. But equally, we would expect them to manifest these qualities in different ways. Some will do it with flair, some with dedication to detail, some with real empathy for users, some with quantitative precision, some with wit and humour and others through the analytic application of rules. And some will do it in ways that we have never dreamed of. Long may it last.

The challenge of differentiation therefore is to retain a balance between, on one hand, *respecting individual differences* and, on the other, holding a clear view about the essential qualities children need if they are to develop technological capability.

Discerning different approaches to Design and Technology tasks

The APU survey provided us with 20 000 pieces of work, two from each of 10 000 pupils. This represents an invaluable repository of pupils' responses to Design and Technology tasks; indeed it is unique in the world. For the two years following the survey, we examined and analysed the work, looking for ways to understand and interpret the different approaches that were employed by pupils. Our approach to this task was subsequently outlined in our final report.

> We felt justified in approaching the assessment function by developing a whole view of capability. Our experience of assessment in design and technology (in a variety of forms) led us to the conviction that it is often easier to identify a high quality piece of design work than it is to say in detail *why* it is high quality. Precisely because of the integrated nature of the activity and the complex interactions of the various aspects of it, holistic assessments of excellence – which allow us to take these interactions into account – have been far more commonplace in design and technology than in many other, more analytic, areas of the curriculum.
>
> (Kimbell *et al.,* 1991)

We recognised, however, that holistic assessment has limitations. It is no good saying, 'This is good, but I don't know why' especially if the purpose of the work is to help teachers to develop their understanding of the multitudinous forms in which pupil capability manifests itself. Accordingly, we devised a means of getting inside the holistic mark to the central traits of good (and poor) performance. But the trick was to do this *without defining these traits too rigidly in advance.*

The strategy was adapted from a well-established technique in experimental psychology.

> It started with us (the research team) in association with a wider group of teachers awarding holistic marks to a range of work from a range of tests in a range of contexts. Our reliability studies gave us confidence in our (and the teachers) ability to agree on this holistic mark. We then took two scripts – a high scorer (a) and a low scorer (b) – and listed all the things that (a) contained that (b) did not, and all the things that (b) had that (a) did not. This was done by looking for questions (does this script do/contain . . . ?') to which one could answer *yes* for one script and *no* for the other. *When we found such*

> *a question we had identified a discriminator of capability.*
>
> (Kimbell *et al.,* 1991)

With a trained group of markers, this technique was pursued with countless pieces of work, resulting in a very long list of discriminating questions. It then became necessary to see to what extent they could be grouped and prioritised.

> We coined the expression 'fingerprinting' the scripts because, like a fingerprint, each script was unique, but by building up a list of discriminating yes's and no's it became possible to describe the uniqueness in any particular script. Moreover, as computers are adept at handling such simple (binary) data, it became possible to ask the computer to generalise these descriptors by selecting all high scorers and printing out the discriminating characteristics that they *did* contain and those that they *did not* contain.
>
> (Kimbell *et al.,* 1991)

Whilst the holistic mark enabled us to *value* a piece of work, the yes/no responses provided a composite *description* of it. We subsequently grouped these descriptors under three headings, reflecting the three dimensions of the model we used to describe the activity in the first place.

These three dimensions of capability are derived from the model shown on p. 24 in Chapter 2 and are described as *reflective qualities*, *active qualities* and the *appraisal qualities* that link the two together.

> These three qualities emerged in pupils' work as the abilities to:
>
> - identify the *issues* in the task;
> - develop *proposals* for a solution;
> - *appraise* the developing proposal in the light of the issues.
>
> Pupils' work does not always have all three components, and seldom are they present in the same proportion.
>
> (Kimbell *et al.,* 1991)

We developed a schematic representation of these three existing in an iterative relationship, and were able to use it to demonstrate some very significant differences in the characteristic ways in which pupils tackle tasks.

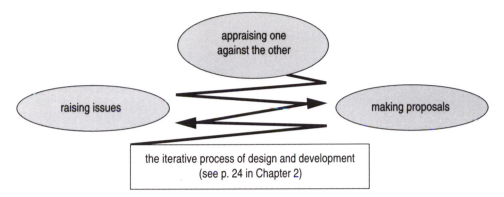

Fig. 6.1 The iterative process of design and development

Three contrasted styles of working

As we pointed out in the APU report, the three clusters shown in Fig. 6.1 that make up capability are seldom present in the same proportion, or even in any kind of balance. It is commonly the case that pupils' capability is skewed towards one cluster and away from another. We can illustrate this phenomenon through three examples of pupils' work from the APU project.

In one case the task involved pupils designing a cooking timer for use by the elderly, and based on a simple twisting concept. The response (see Fig. 6.2) is almost entirely written (in text) and shows a highly developed understanding of the user issues that need taking account of in this situation, but the pupil is trapped by an apparent *inability to make concrete proposals* that address these issues. As virtually no design ideas emerge, there is nothing for the pupil to appraise, and the pupil is left with nothing to do but re-state the issues already identified and the design criteria given through the task brief. This is a classic example of unbalanced capability skewed heavily towards the reflective identification of, and discussion of, *design issues*; but devoid of creative action.

A second example (see Fig. 6.3) shows a very contrasted imbalance. In this case, the task involved the design of a child's moving toy. The response shows a clear understanding of how one might create moving parts on a clown face, using mechanisms of several kinds. But the idea ends up exactly where it started. There is no growth and development at all because the pupil is *unable to pull out the issues* that might be important to improving it. As the task unfolds, the pupil is left with little to do except redraw it more prettily in almost exactly the same form as the original idea, but embroidered somewhat through the use of colour.

These two unbalanced responses represent opposite styles of working, and the level of response that becomes possible when the two are combined in the same pupil is apparent from our third example (see Fig. 6.4). In this case the task involved the design of a stackable plant holder. Here the pupil immediately jumps in with some extremely simple ideas but subjects them to some very thoughtful questions about how the user might interact with the product. The result is that the proposals grow in diversity and in detail, ending up at a far higher level of sophistication than the point at which they started. This is, as we said in the final APU report 'a classic example of highly developed design and technological capability'.

As we have suggested above, it is possible to characterise these three pieces of work as different outcomes. The differences in the outcomes reflect the balance that exists in the three dimensions of

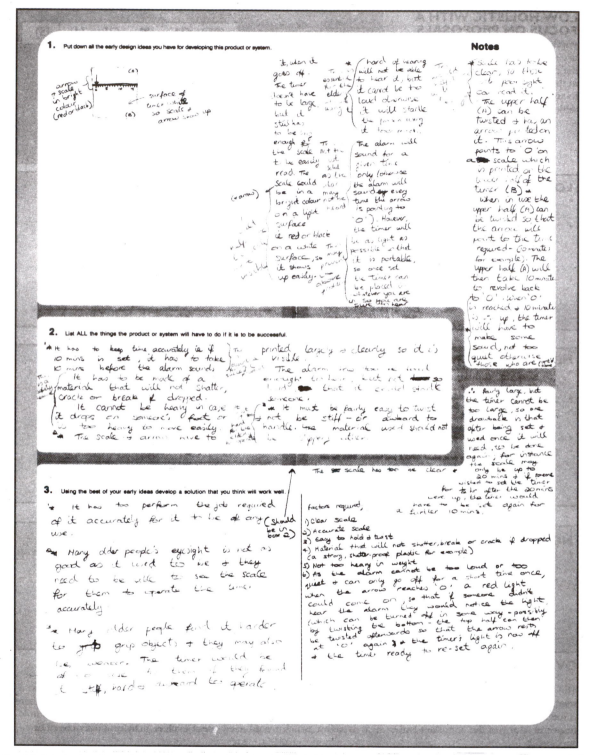

Fig. 6.2 Work showing a heavy skew towards reflection (APU report p. 175)

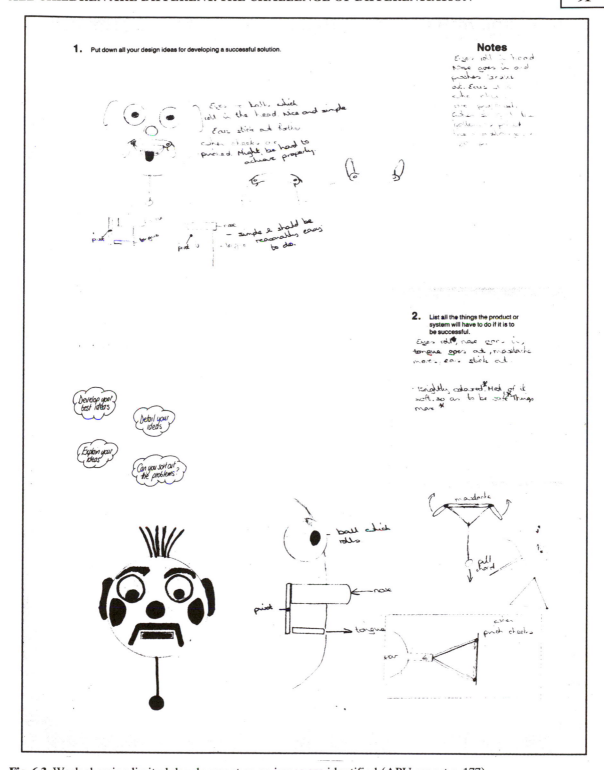

Fig. 6.3 Work showing limited development as no issues are identified (APU report p. 177)

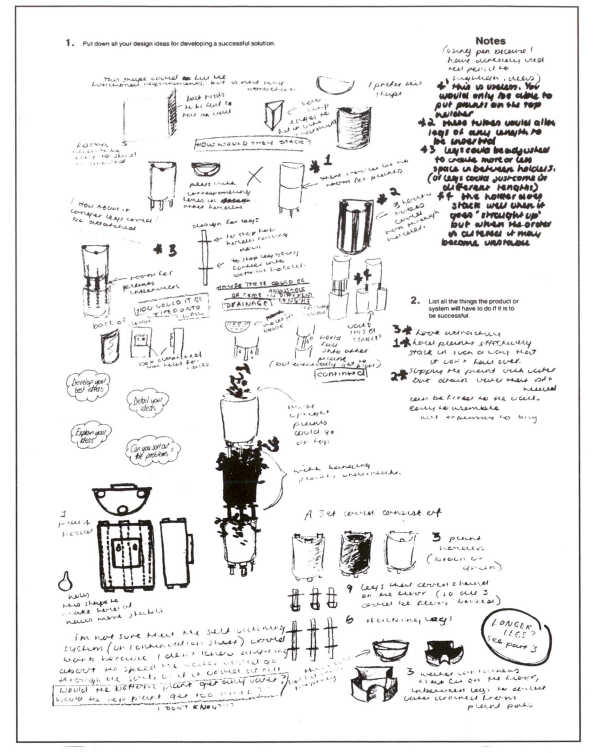

Fig. 6.4 An example of highly developed, balanced Design and Technology capability (APU report p. 181)

our framework for describing Design and Technology; identifying *issues*, making *proposals*, and *appraising* one against the other.

They are, however, different in other ways too, and these further differences concern the *input* conditions that led to these differences of performance; i.e.

- there were differences in the context of the tasks;
- there were differences in the definition (and detail) of the tasks;
- there were differences in the structure of the activities.

These, as will be recalled from Chapter 4, are all input differences that have a marked impact on the difficulty of the task. They can therefore be expected to have a significant impact on the subsequent output/performance from different subgroups of pupils. This brings us to the final measure of difference, the pupils themselves.

In the remainder of this chapter, we shall explore the interactions between some of these groups of variables (Fig. 6.5). The *task input variables* are the three outlined above: the context, the task definition and the structure that the teacher imposes on the activity. We shall explore these differences in

terms of two broad *pupil variables* viz. gender groups and 'general ability' groups.

We are dealing here with multiple variables; mixing input variables for the task/activity with person variables, and the whole mixture might be summarised in the following terms. Input variables create performance differences which can be analysed in terms of gender and ability groupings. The messages that one might seek to draw from the analyses are made more difficult to unpick by virtue of the interaction of the two groupings.

As individuals, we all belong to groups of many different kinds and these groupings reflect – to some degree – our propensity to behave in certain ways in response to certain situations or conditions. They affect the ways in which we experience the world and hence to some degree they affect our learning.

One of the purposes of APU surveys was to analyse performance against a number of standard variables – including gender and 'general ability' – in order to derive messages that might help teachers to understand and respond to the strengths and weaknesses in the performance of pupils in these groups. We recognise that the 'general ability' category is a difficult one to define, but the APU approach to this was relatively easy for

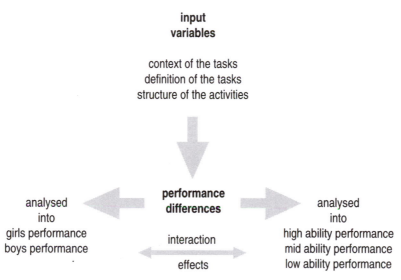

Fig. 6.5 Interaction of variables

schools to operate (at least for age 15 pupils) and has provided some valuable insights into performance differences.[1]

The different performance of gender and ability groups

The following analyses represent a brief summary of the full findings that appear in section 15 of the APU final report (Kimbell *et al.,* 1991).

- Generally, girls do better than boys in the more reflective areas of work, for example in identifying the *issues* that underlie tasks, in empathising with *users* and *evaluating* products and systems in terms of how well they might perform for the user.
- Generally, boys do somewhat better than girls in more active areas of work, for example in *generating* ideas and in *modelling* them through to working solutions.
- Further evidence to support these two findings has come from the Key Stage 1 research on Non-Statutory SATs in Technology.[2] Girls were found 'to have outperformed [boys] in all but AT3 [making]' (Stables *et al.,* 1991).
- Generally, girls do better than boys when tasks are loosely defined (allowing scope for personalising it) and boys do better than girls when the task is tightly defined (when it is clear what has to be done).
- Where these effects work together they compound each other:
    tightly defined and dominantly active pieces of work will enormously favour boys;
    loosely defined and dominantly reflective activities will enormously favour girls.
- Where these effects are set against each other, they tend to cancel each other out:
    tightly defined and dominantly reflective pieces of work favour no-one;
    loosely defined and dominantly active pieces of work favour no-one.
- Generally, the lower the ability group (girls and boys), the more exaggerated are the effects described above:

    low-ability boys do relatively well in modelling and poorly at evaluating;
    low-ability girls do relatively well at evaluating but do very poorly indeed at modelling.
- There is generally less effect on the high-ability groups.
- Generally the low-ability groups (girls and boys) do better in clearly structured activities. For high-ability groups, the structure of the activity is less significant.
- The context effect is such that girls out-perform boys in contexts focused on *people*, and there is a tendency for boys to do better in *technical/industrial* contexts. The context of *environments* would seem to be largely gender neutral.
- Again, these effects are more marked when the sample is split by ability with low-ability girls really under performing in industrial contexts and low-ability boys in people-centred contexts.

We concluded this section of the APU report by stating what should be obvious:

- Bear in mind that the *context* of the task will affect how it is received and understood, and particularly so in the case of lower ability pupils.
- Bear in mind that the details of the *specific task* and particularly its 'openness' or 'closedness', will affect the ease with which pupils can deal with the task and particularly so in the case of lower ability pupils.
- Bear in mind that the *procedural structure* that you build into the activity, particularly the active/reflective balance and the 'tightness' or 'looseness' of the structure, will seriously affect pupils' ability to get to grips with the task and particularly so in the case of lower ability pupils.

(Kimbell *et al.,* 1991)

Counteracting 'characteristic' weaknesses

When considering what teachers might do to counteract these tendencies, we looked first at the problem of helping pupils to improve their *reflective* (issues centred) performance.

It was evident that boys are more able to get to

grips with reflective aspects of capability when they are practically engaged in developing a solution, and especially so when they are able to do this through more practical modelling activities. Girls would appear to be much more able to empathise with users (and identify the critical issues that need to be dealt with) without the benefit of such practical engagement.

We then looked at the problem of helping pupils to improve their *active* (proposal-centred) performance, and here again we should not be surprised with what we find. Girls are better able to get to grips with active modelling and idea development *when they can see it through a people-centred focus* (e.g. in a 'people-rich' context). Boys appear to be more prepared to engage in such practical activity without this people focus; indeed they appear to relish it in any setting. Technical tinkering is a male characteristic.

Predicting performance on hypothetical tasks

In combination, these data suggest that we might predict the performance of sub-groups just by knowing the details of the task and how it is structured. How, for example, might we expect the sub-groups to perform with the following kind of task?

- a focused technical task (using Lego)
- that is loosely structured
- to develop a mechanism that gives output motion B from input motion A.

One could predict that the boys, even low-ability boys, will engage in the task with little difficulty. The low-ability boys will find the lack of structure a problem, but will overcome it because the task is in their comfort zone of technical tinkering. The girls might be expected to find it more difficult to engage with the task unless there is something that gives it some explicit purpose. The low-ability girls in particular will find the lack of structure very threatening, since the task is well outside their comfort zone. One might therefore predict that (if the APU data are to be believed) we would end up with the performance spectrum for this kind of task illustrated in Fig. 6.6.

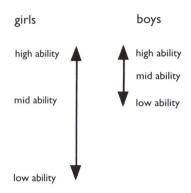

Fig. 6.6 Performance spectrum for a focused technical task

Further evidence of these reactions comes from a recent study by Ross and Browne (1993) focussed on *Girls as Constructors in the Early Years*. Specifically in Key Stage 1 classrooms, they highlight the extent of the gender differentiation in reaction to tasks.

> All the boys were enthusiastic about the Lego table. Few girls said they played with it, and then only when pressed, and some girls said they didn't like Lego at all. The boys used Lego in much more sophisticated ways making more use of it as a medium, exploiting its three-dimensional properties (e.g. balance, capacity for movement, potential for complexity of configuration).
>
> With the boys there was a constant process of making and remaking Lego constructions. They played with their constructions on their own. Girls often brought 'play people' over to the table to use as dolls, and used their Lego houses as a prop for social play about families at home. Girls played 'house' with each other, constantly interacting as an intrinsic part of play.
>
> (Ross and Browne, 1993)

So what might we expect from a quite different kind of task ? What if it were to be as follows:

- a focused evaluative task
- to discover the 'ideal' can-opener that might be used by pupils' elderly grandparents
- tightly structured through a series of sub-tasks.

The data would suggest that, because the focus of the task is reflective and human centred, it will

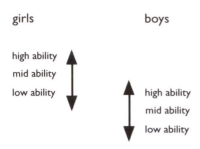

Fig. 6.7 Performance spectrum for a focused evaluative task

favour girls. The fact that it is tightly structured through a series of pre-planned sub-tasks will ensure that low ability girls will also perform well. By contrast, the boys might be expected not to engage at the same level. There is not much *to do* in active design terms, and the premium on talking to people and thinking about *their* priorities and concerns places the task outside boys' comfort zone. The tight structure, however, will tend to support all pupils. One might therefore predict that (if the APU data are to be believed) we would end up with a much more tightly distributed spectrum of performance for this kind of task (Fig. 6.7). The reader will no doubt have a view about this. Do these predictions match your experience?

As we pointed out in Chapter 1, it is quite possible to place both sinister and constructive interpretations on our understandings of the multiple effects that these contrasted activities might provoke.

> . . . it would be possible – given an understanding of the nature of these effects – to design activities deliberately to favour any particular nominated group. More positively, it would also appear to be possible to design activities that largely eliminate bias or at least balance one sort of bias with another.
>
> (Kimbell *et al.,* 1991)

However, there is a further vital matter to be recognised here. The analysis of APU data might allow us to predict these levels of performance from different sub-groups – *but the findings are based on data that encapsulate the reality of what*

we have seen rather than what one would want to see. It is not good enough to say that girls prefer X so give them lots of it, or that boys do better with Y so we must always do that.

Building from strength to overcome weakness

From the APU data, it is a fact that different people will approach the same task in different ways. But it does not follow that the approaches are all equally desirable, or that the learner would not make more progress if they were equipped to do it differently. Education is not about supporting the status quo, but is rather about enriching what pupils are able to do. A pupil who can take design ideas forward through discussion, drawing, making, research into materials, looking critically at other solutions and so on, inevitably has more resources at their disposal than the pupil who is superb at working things out on their own, on paper, but who is effectively paralysed if required to work in any other way.

We need consciously to work *from* the learner's strength rather than *on* the learner's strengths. This point is central to learning at any point in the 5–16 age range. Ross and Browne (1993) reiterate the point in relation to girls' performance in Key Stage 1:

> One way to help girls to overcome their reluctance to engage in constructional play is to begin with a familiar area and move gradually into 'unfamiliar territory' . . . starting where girls are confident, competent and interested.
>
> (Ross and Browne, 1993)

So a study of mechanisms starts with the need to make Easter cards (that are subsequently made to 'pop-up'), or with drawings of their favourite animal (that can subsequently have articulating limbs). The same messages apply for the boys of course, though the challenge here is to encourage them into empathising with the users (e.g. of their Lego constructions) through their enthusiasm for developing working mechanical constructions.

The differences in the strengths and weaknesses of the gender groups, therefore, can be seen to relate both to the substance of the task (e.g.

mechanical constructions) and to the procedural arrangements that the teacher builds around it (e.g. the tightness of the task and its sub-structure). A good general rule would be that it is desirable for all pupils to become equally familiar and comfortable with all kinds of content and procedural arrangements. In this way, pupils will hopefully develop a range of approaches to their work that can be deployed appropriately in response to the different demands that exist in different tasks.

This issue is at its most evident in the critical area of design development, generating ideas, exploring them and taking them forward to workable solutions.

Design development provides some interesting examples

As we pointed out in Chapter 1, one of the principal claims for Design and Technology in the curriculum is that it provides pupils with 'a concrete lever that can expose and get a purchase on their thought processes'. We need to recognise, however, that there are a multitude of ways to do this and not just one way.

Differentiated approaches to design development require and thrive on this diversity. Ideas can be developed in any number of ways:

- through discussion;
- through concrete modelling (with paper/Lego/foam/'Plasticine', etc.);
- through drawing;
- through taking photographs;
- through observation and recording;
- through taking things apart to see how they work; and
- reassembling them to see how all the bits interlock.

It is one of the commonest and most serious misconceptions in Design and Technology that design development can only happen on paper with a pencil. For many children, and particularly for boys, such a requirement places quite insuperable barriers to the development of ideas. It requires

skill and confidence with a pencil and sophisticated graphic concepts such as the ability to represent real 3D objects on a flat piece of paper. No one could sensibly oppose the progressive development of such skills and concepts, but to make them a preliminary requirement of any designing activity is to place unreal and unhelpful barriers between children and the development of their ideas. The *concreteness* of Design and Technology is a major part of its appeal and the source of its power to support children's learning.

A group of our first-year undergraduate students has just completed a project to design and make a range of banners to hang in the entrance to the Design Studies department at Goldsmiths College. Within minutes of being set the task, different working methods became obvious. Some students talked, some sketched on their own, some sketched and talked, and some left the studio to go to look at the entrance space, soon to return for tape measures, scissors, 'Sellotape' and newsprint to mock-up some early ideas on the spot. This latter group had found that they needed to work things out *in situ* in order to explore and develop their ideas. Each approach was appropriate and helped to move things forward.

In our classroom observations in the UTA study, we have witnessed countless examples of pupils struggling to work things out on paper when the task cried out for a greater variety of approaches with some concrete exploration. One such case was of a 12-year old, who was very confused about the measurements and dimensions needed to produce a 'cutting list' for a wooden box. She sat for a long time looking (helplessly) at her paper and eventually began to talk it through with her friend. Subsequently, aided by a ruler and a different type of box (made by a previous pupil) she saw what she needed to do and moved on confidently to complete her own list.

In another case a younger pupil was trying to design and make shoes fit for royalty. He was faced with a similar problem, and had great difficulty working out the detail of his shoe on paper. His project really got going when he realised that he might make a paper shoe for his own foot. Armed with paper, scissors, a tape measure and some pins he set about making the shoe, and in an hour of intense and concentrated 'trial and error', he produced a paper model of his shoe. It fitted his foot and could be taken on and off. As a result of this endeavour he also had a pretty reasonable idea of how to set about making the real thing. This is a classic example of thinking being liberated and extended by the kinds of concrete modelling that is such an essential feature of Design and Technology.

The point here is that differentiated approaches to design development require that we recognise the diversity of ways of doing it. Some pupils will naturally favour one strategy over another and part of our role as teachers is to enable them to expand their tool-kit of approaches. The very worst thing we can do is to insist that there is only one way to do it.

The general issue of differentiation

This specific point about design development is representative of the more general point about differentiation. If a pupil's current tool box of strategies is not adequate to help them work out some aspect of their design, then it needs to be enriched. The issues that we have drawn attention to above about the 'typical' strengths and weaknesses of gender or ability groups are therefore not something about which we just shrug our shoulders. We should not accept these strengths and weaknesses as anything other than starting points. But in order to build from them, we need first to recognise that they are (probably) there and only then can we work *from* existing strengths and *towards* creating new ones.[3]

Just because we are aware that boys may have more difficulty than girls in reflecting on their work, does not mean that they should not be encouraged to do it. Quite the reverse in fact, since to avoid reflecting would be to deny them the opportunity to develop this critical dimension of capability. Our understanding of the issue merely helps us to provide appropriate support for them as they tackle the challenge.

As we pointed out at the start of this chapter, differentiation is about being aware of the differences between learners and taking account of these differences in our teaching. This issue ultimately amounts to an individualised curriculum, but for reasons of manageability we have dealt with it here as a matter of understanding the predilections of sub-groups of pupils. The evidence that we have drawn upon has been predominantly the APU survey data, and we have every reason to believe in its veracity. Nothing that we have observed in our recent extensive classroom observations at Key Stages 1, 2, 3 and 4 has suggested that the typical reactions of sub-groups are anything other than as we have outlined here. Nor do we believe that teachers will be surprised by the tendencies to which we have drawn attention.

The trick is to see these typical differences in approach as an opportunity, rather than as a problem.

SUMMARY

- Differentation is about being aware of the differences between learners and taking account of these differences in our teaching. It is therefore ultimately an equal opportunity issue.
- Following the APU survey, we sought to unpick thousands of pupil responses (each one unique) in ways that are helpful to our understanding of capability. We developed an approach that we called 'fingerprinting' in which unique performances can be analysed into component parts.
- The approach resulted in our being able to highlight three characteristic ways in which pupils work:
    reflective
    active
    integrated
These are exemplified.
- Whatever groups we belong to (e.g. gender groups) they affect the way we experience the world and hence they affect our style of working and our learning.
- It is illuminating therefore to analyse different styles of working in terms of different sub-groups of pupils (e.g. girls/boys).
- We use the APU data to detail a series of

characteristic performance differences between gender and ability groups.

- We then use these data to predict the outcome performance of different groups in response to two hypothetical and contrasted tasks.
- It is important to remember that the point of understanding these differences is not so that we can constantly allow boys (for example) to do what boys are good at, but rather to show how we can work *from* these strengths to improve other areas where they are less strong.

Notes

1 The approach was to ask schools to place pupils (identified to APU only by date of birth) on a six-point scale based on their predicted GCSE performance. The top of the scale would be 'will get 5 or more GCSEs with at least 3 at grade A' and the bottom of the scale is 'will not be entered for any public examinations'. This six-point scale is subsequently reduced to a three-point scale of 'above average', 'average' and 'below average' ability groups. This approach has been used for all APU surveys, in Science, Maths and Languages, as well as in Design and Technology.

2 The research project developing Non-Statutory Key Stage 1 SATs in technology was directed by Kay Stables at Goldsmiths University of London between 1990 and 1992.

3 It is important to remember that the general trends we have observed in the performance of gender and ability groups are nothing more than that – they indicate the *tendency* of individuals to respond in one way as opposed to another. Not all girls, boys and high/mid/low-ability pupils will behave 'true to type'. But there is good reason to believe that they will tend towards the strengths and weaknesses outlined here.

CHAPTER 7

Transition: exploring the Year 6–7 boundary

Introduction

We have indicated in earlier chapters some of the discontinuities that exist in the Design and Technology experience of pupils in schools, and by far the most extreme example of this is between Year 6 and Year 7, the transition from primary to secondary school. In this chapter we shall explore these differences more fully and highlight the factors that polarise the experience. We also make a series of suggestions for creating a more progressive and coherent experience for children.

We shall attempt first of all to characterise the experience of these two years by looking at four case studies of children in the UTA study (Chapter 1); two at Year 6 and two at Year 7. We shall then examine further data from the study, to highlight several features that set the tone for the teaching–learning partnership.

Twenty-eight of the 80 children in the UTA study are clustered around the Year 6–Year 7 boundary, and for all the children surveyed, this was the point of transfer from primary to secondary schooling. The 28 children were monitored (four at a time) pursuing the following range of projects:

Year 6
- designing and making a museum exhibition for the 21st century;
- designing accessories for wearing on bicycles;
- designing and making simple moving vehicles;
- making a model English galleon.

Year 7
- designing and making an action sports trophy;
- designing and making a plant-watering sensor;
- designing and making a mechanical toy.

These projects are not untypical of projects seen elsewhere in Year 6 and Year 7 classrooms.

In the UTA study, the Year 6 projects typically (but not exclusively) derived from other thematic classroom work and had a problem-solving focus. The Year 7 projects were derived from tightly defined design briefs and having a knowledge and skills-development focus. They were all concerned with designing and making but there were significant differences in the ways in which the projects were structured and the roles taken by the teachers.

Characteristic projects

Four projects are outlined below, two from Year 6 and two from Year 7. The outlines describe the nature of the work the children were involved in and are intended to provide a concrete reference point for the discussions that follow. All the projects were successful in their own terms; they motivated the children and promoted learning in Design and Technology.

Project 1: Primary Years 5 and 6 – designing and making a museum exhibition for the 21st century

Set in a mixed Year 5–Year 6 classroom, this project was developed out of a history topic in which the children were considering ways in which the past can be presented and interpreted through primary and secondary source materials. Four children in the class worked together to produce a museum-type exhibition, that could be displayed a hundred years from now and that would tell people what life was like for children living in the 1990s. The children were given no time limit for the work or any specific restrictions on resources, other than that the exhibition had to be mounted somewhere within the classroom and utilise resources generally available within the school. Initial discussions, instigated by the teacher, involved the children thinking through aspects of their lives that would typify the era and for each aspect some ideas about how these could be presented effectively. Their ideas ranged from collecting artefacts such as coins, food packages and videos to creating their own newspaper with articles about how children spend their leisure time, what school life is like and so on. At the heart of their ideas was the construction of two models, one of a house to show the sorts of things (in particular the technology) that people have in their homes and one of a supermarket to show how people shop for food and what they eat and drink. The main efforts of the children went into producing these models which were created by them working initially as a foursome to do some early thinking and thereafter mostly in pairs to work out the detail. The ideas were developed almost entirely through working directly with materials. Very little was drawn or written for either planning or recording purposes and the children threw away all paper work as soon as they perceived they had no further need for it (despite having been asked not to!). The teacher operated mainly as progress chaser, sitting down with the group every so often to review work in progress and help them plan the next stages. Very occasionally she stepped in to provide specific guidance, typically if one of the

children was having difficulty in an aspect of manufacture. The focus of her interactions with the children were predominantly of a questioning nature, aimed at encouraging them to think through the detail and consequences of their ideas for themselves.

Project 1
- Children working as a group;
- task – open-ended and negotiable;
- task – from class topic, fed into other classwork;
- materials – that which is available in school;
- constructions – as appropriate/available;
- time – no limit (actual time 20 hours);
- input predominantly supportive.

Project 2: Primary Year 6 – designing and making moving vehicles

This Year 6 project was seen as a separate Design and Technology project and did not draw directly on a class topic. The project took place on a specific afternoon for six consecutive weeks and involved the whole class working at the same time in their general classroom. The pupils worked individually, although there was a great deal of informal collaboration, as pupils discussed and supported each other in their work. The task was to create a moving vehicle, but it was up to the children to identify their own focus. Of the five children observed, one chose to make a wind-powered boat, one a battery-powered buggy, one a 'windmill on wheels', one a crane and one a fairground carousel. The teacher instigated initial discussions, made specific skills inputs to the whole class, drew them together to review their progress and provided ongoing individual support. She also encouraged the children to use a diary for reviewing and planning their progress in each session, working under the headings of 'What am I doing today?', 'What have I done today?' and 'Am I pleased?' As with the previous teacher, the focus of interactions with the children was predominantly of a questioning nature, providing them with a 'pause for thought', but throwing the responsibility for the thinking and decision-making directly onto the children themselves. The outcomes were seen as models and the children were visibly involved in problem-solving as they pursued their ideas and got the models to work. This involved working in part through drawings, but mainly through the materials themselves. They utilised previous knowledge and skill, discussed with the teacher and their friends and engaged in a good deal of trial and error, all finally producing a 'moving vehicle'.

Project 2
- Children working individually, but in a collaborative manner;
- task – specific constraints, but thereafter open-ended (all produced different outcomes);
- task-specific to 'timetabled' Design and Technology;
- materials – that which is available in school;
- constructions – as appropriate/available;
- time – 6 weeks (actual time 6.5 hours);
- low proportion of teacher input.

Summary: Projects 1 and 2
- Teachers primary trained;
- workspace – general classroom;
- materials/processes – not specified;
- designing – carried out mainly through materials;
- teacher's main role – progress chaser.

Project 3: Secondary Year 7 – designing and making a plant-watering sensor

This secondary project was set in an all-girls' comprehensive school. It took place towards the end of Year 7 and was aimed at introducing the children to electronics and some basic woodworking skills, and at further developing their design skills. The lessons were timetabled on a weekly basis, each lasting just over 2 hours. Six lessons were allocated to completing the project. The work took place in a specialist Design and Technology room. The project built on a common approach (a standard project folder) used by all teachers in the department. The children worked to a checklist of headings (title, brainstorm, design brief, analysis, guide to making a PCB, six designs for a suitable box,

cutting list, three designs for getting battery out, examples of LEDs, safe use of soldering iron, circuit diagram, evaluation). The children worked individually on their own projects, but were encouraged to work in small groups for initial brainstorming. These groups often remained as informal discussion and 'sounding board' mechanisms, although some children only worked collaboratively when asked to. The children were given the option of designing anything that utilised a water-sensing unit, and a range of uses were identified through a whole class brainstorming session. However, all children chose to make a plant-watering sensor. Teacher inputs were both to the whole class and to individuals and were predominantly of a technical nature. The incidence of teacher input was average for Key Stage 3 (although high overall) and the interactions were either transmitting information or questioning pupils to encourage their own thinking.

Project 3
- Children working individually, but groups used for discussion;
- task-specific brief set (all produced similar outcomes);
- materials – specified by teacher;
- manufacturing processes all common;
- task time specific to 'timetabled' Design and Technology;
- time – 6 weeks (actual time 13 hours and homework);
- high proportion of teacher input – predominantly directive, but less so than others at Key Stage 3.

Project 4: Secondary Year 7 – designing and making an action sports trophy

This project was set in a Year 7 secondary workshop in a non-selective, grant-maintained boys' school. For the teacher, there were several aims to the project: to introduce the children to working in resistant-materials workshops; to work with metals; and to work with a design process. The brief required them to design and make a trophy which had as its centre point a stylised forged figure using metal rod and strip. The children had to make, shape and position the figure on a base in relation to their chosen sport. The base of the trophy and any further attachments to it, such as balls, nets and so on were left for them to design as they saw fit. The project had to be completed within a seven-week period, each week providing two workshop sessions of 70 minutes and a theory lesson of 35 minutes. The children worked individually, although taking a great deal of interest in each other's work and in previous work done from the same brief. They all produced a trophy and a design folder to accompany it. The role of the teacher was predominantly that of instructor/facilitator providing specific and detailed guidance on how to use tools and equipment, represent ideas, plan on paper and work through specific aspects of a design process. The teacher made the highest number of inputs of any teacher observed in any of the UTA studies. There was a mixture of those aimed at direct transmission of information and those of a questioning nature aimed at encouraging the children to think through the issues involved in their designing and making.

Project 4
- Children working individually;
- task-specific brief set;
- materials – specified by teacher;
- manufacturing processes all common;
- task time – specific to 'timetabled' Design and Technology;
- time – 7 weeks (actual time 17 hours + homework);
- high proportion of teacher input – predominantly directive.

Summary: Projects 3 and 4
- Teachers secondary trained;
- workspace – Design and Technology workshop;
- materials/processes – largely fixed;
- designing – done in advance on paper;
- teacher's main role – instructor/facilitator.

It is worth examining the summary statements again, since the contrasts between the two models of teaching and learning Design and Technology are so marked that they must appear (to the children) to be completely different things. It is perhaps worth recalling that the distance between these two utterly different models of practice is one summer holiday.

The four projects provide strongly contrasted experiences and much of the contrast centres on the issue of *certainty*. At one extreme we have primary pupils engaged upon something the outcome of which is uncertain both to the pupils and (to a large extent) to their teacher. Furthermore, the procedure by which they get to the outcome is also uncertain since 'the next step' typically evolves from discussions about where we are now. The teacher is not a specialist in Design and Technology and typically does not therefore know all the answers about how to do things. Primary children are expected to operate increasingly autonomously in a context of uncertainty.

By contrast, the Year 7 teachers organised their projects into carefully controlled steps towards known outcomes. The teacher is a specialist and carefully builds in the need to teach specific skills and knowledge – but in a way that makes these needs arise 'naturally' through the project. The children see the teacher as knowing all the answers. The process skills – for developing and recording ideas – are equally controlled by the teacher through the folder format. There is very little freedom of action to pupils (which is why all the outcomes are so similar) and there is equally little uncertainty.

- *Uncertainty, and 'working things out' typify Year 6 projects.*
- *Certainty and 'right answers' typify Year 7 projects.*

Ironically, these differences result in very different levels of confidence and personal autonomy being displayed by the pupils. The Year 6 pupils become increasingly independent as they realise that they can make decisions and make things happen as a result. The Year 7 pupils, on the other hand, rapidly become dependent upon their teacher (and spend

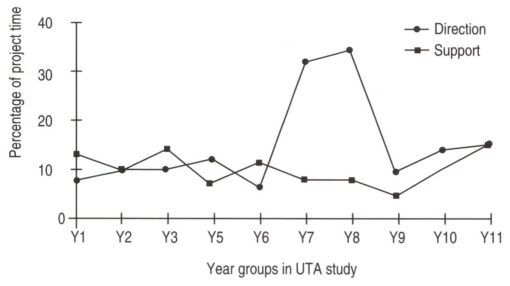

Fig. 7.1 Teachers supporting children - or directing children

long periods standing in queues waiting for advice) not wanting 'to make a mistake'. In a strange way therefore, dealing with uncertainty breeds confidence whilst constant certainty breeds dependence.

In the following pages we analyse several features of the teaching and learning exchange that contribute to this general condition.

Teachers directing and teachers supporting

We distinguished earlier between teacher–pupil interactions that are *directive* (telling them what they have to do) or are *supportive* (responding to the child's own needs or ideas) (see p. 44). We have recorded these data for every five minutes of every project observed in the UTA study. We can therefore look at the incidence of 'directing' and 'supporting' through the life of projects and express it as a percentage of the project time. Fig. 7.1 has already been reproduced in Chapter 4, but it is worth using again since it shows very clearly what happens to each year group (there is no data for Year 4) in terms of their exposure to 'direction' and 'support'.

With the exception of Years 7 and 8, the differences are marginal, with teachers directing and supporting in equal amounts and at around 10 per cent of project time. There is very little change from year to year. But in the first two years of Key Stage 3 there is a dramatic rise in teacher direction up to 35 per cent of project time.

The use of this direction is largely to enhance pupils' understandings and skills with materials and processes. As a result there is no doubt that the pupils do learn a wide variety of making processes that they would otherwise not experience. Key Stage 3 is typically very rich in new material experiences, with food, metals, fabrics and plastics, and we should not underestimate the thrill that these new experiences can provide. But the tight framework within which it is received very often prevents pupils from exploring what is possible with these new processes, and Fig. 7.2 provides an interesting example of this.

A Year 8 group was being introduced to enamelling as part of a six-week brooch project. One pupil's output is shown in Fig. 7.2. The brooch is the (somewhat limited) star-shaped piece, fixed to a brooch-pin. As the project developed, he became less and less concerned with the outcome being a piece of jewellery and more and more fascinated simply by what it was possible to do with enamel as

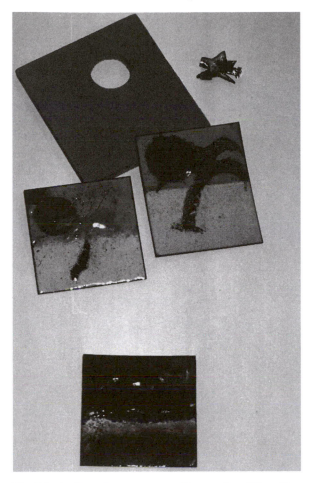

Fig. 7.2 An introductory enamelling project (Year 8)

Different use of time

We have made the point earlier, and particularly in Chapter 3, about the importance of 'hands on' experience. Design and Technology teachers would typically place great faith in the old adage 'I hear and I forget; I see and I remember; I do and I understand'. The idea of concrete learning is deeply embedded in the folklore. Yet the palpable consequence of the teaching style typically adopted in Year 7 is to reduce significantly pupils' practical engagement with the task, despite the fact that the projects are longer.

Total time on projects rises (on average) from around 10 to 12 hours, but the exceptional rise in the amount of 'listening' time has the effect of reducing the 'working' time on the project, see Fig. 7.3. Interestingly if we look at the data across the whole 5–16 spectrum, the boundary between Key Stages 2 and 3 marks the only point at which this regression occurs. This raises some interesting questions about how teachers see their roles. Teachers at all four key stages would no doubt claim that they are teaching, i.e. introducing children to new skills and knowledge, new working practices and procedures, developing their attitudes and values. However, the instructional nature of the first two years of Key Stage 3 sets it completely apart from the rest of pupils experience.

a decorative medium. He produced – with growing enthusiasm and delight – a series of enamel panels that were not only beautiful in their own right but moreover explored the technicalities of overlaying (and masking) layers of enamel. He was totally hooked on – and thrilled by – the process. But one would never have thought so by looking at his brooch, which was the ostensible 'outcome' of his project.

The directional nature of so many Year 7 projects, and the lack of time that this allows for exploration of these new (and often fascinating) technologies, raises some interesting questions about the use of time in projects.

What are teachers doing?

Given the contrast outlined above in the amount of direction applied by teachers, it would clearly be instructive to examine the nature of what is being passed on during this intensive teacher direction and to examine the extent to which it varies from what is being passed on at other points in the 5–16 spectrum.

Our data enable us to examine whether the teacher intervention is linked primarily to *user* issues (e.g. Could the 'action sports trophy' be kept clean easily? Is it comfortable to use? Can it be repaired if it breaks?) or to *making* issues (e.g. how can the figure be fixed to the base? How can I bend

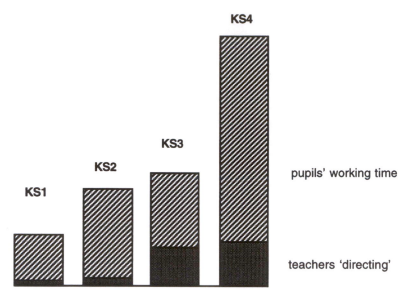

Fig. 7.3 Available working time reduces between Key Stages 1, 2, 3, and 4

the parts of the figure? How can I join the parts together? How do I fix it to the base?).

We can explore these questions both in terms of the 'direction' provided by the teacher and in terms of the more personalised 'support' that they provide to individual pupils. A clear picture emerges (Fig. 7.4). With 'user' issues there is very little change between what teachers are doing in Year 6 and what they see as important in Year 7. This position holds true for 'direction' as well as for 'support'. There is a very small, but insignificant, change and generally teachers will be encouraging

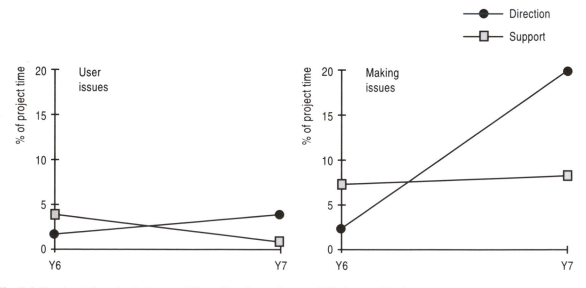

Fig. 7.4 Teacher 'direction' rises rapidly at Key Stage 3, especially for 'making'

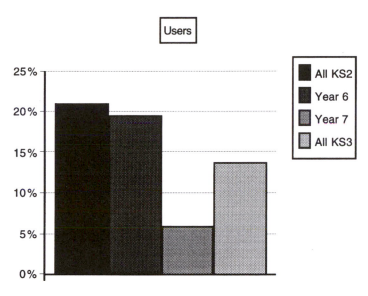

Fig. 7.5 'Users' are far more important at Key Stage 2 than at Key Stage 3, and the Year 6/Year 7 gulf is especially marked

their pupils to think about these concerns for about 3 per cent of project time.

In terms of 'making' issues, there is a very significant change from Year 6 to Year 7 and the focus is clearly on *directing pupils in the techniques of making*. From having been an insignificant concern for Year 6 teachers, it has become a major concern for Year 7 teachers and takes up about 20 per cent of project time.

Children's concern with 'users'

Figure 7.4 indicates the extent to which *teachers* across the Year 6–7 boundary have different priorities about the level of direction and support they provide in relation to user and making issues. However, children's concern is equally interesting for, as we have pointed out in Chapter 2, it is the user that provides the rationale for the technological activity in the first place. Why do we need an action sports trophy unless it is *for* someone?

In terms of children's preoccupations, the issue of how the product interacts with its end-user is considered less in Key Stage 3 than in any other key stage, and Year 7 marks the low point of this concern. Figure 7.5 illustrates the overall positions

of Key Stage 2 and Key Stage 3, and shows the specific difference between Year 6 and Year 7. Whilst Year 6 children regularly consider the relationship between what they are making and how it will be used (approximately 20 per cent of working time), Year 7 children consider this relationship far less frequently (approximately 6 per cent of working time). This has interesting consequences for the evaluation of the outcomes that result. One might imagine that it would be helpful to use a fundamental design criterion like 'fitness for purpose'. But such a criterion only makes sense to pupils if they have explicitly sought to match the product to a real purpose – which typically has to mean a real user.

Moving from tacit to explicit understandings

Some of the differences to which we have drawn attention here reflect differences in the extent to which teachers are seeking to impart explicit understanding (typical of Year 7), or capitalise on tacit understanding (typical of Year 6).

Many of the things that we learn to do (swimming for example) combine these levels of awareness, but at the outset (as a novice swimmer) the tacit understandings will predominate. Getting in the water – feeling the difference in our 'weight', taking our feet off the bottom, splashing our arms, holding our breath – all provide us with experiences that build some tacit understanding about the nature of swimming. Learning to swim is tightly related to the *experience* of swimming and even successful swimmers may be unable (explicitly) to explain how it works. They just know they can do it. This tacit level of operation, through which we survive most of the world, is the source of the old belief that you only really understand something when you try to teach it. To teach it, you need to be able to make it explicit – at least to yourself.

Helping *to make explicit* the things that they are able to achieve tacitly is at the heart of children's learning, and this is just as true in Design and Technology as it is in any other field. Moreover, it appears to be one of the differences between Year 6 and Year 7.

Primary teachers appear to be concerned to allow their pupils to experience technology and to capitalise on their tacit understandings. Through experiential activities (and often using play) primary children develop tacit understandings about all kinds of technological matters: how materials behave; how we can fix things together; how we can make things move. They increasingly make sense of the made world through these tacit understandings. See for example the work of Bruner (1968), Baynes (1986, 1989, 1992a), Stables (1992) and Davies (1994).

Secondary teachers, by contrast, are constantly seeking to make these understandings explicit – so much so that they spend a lot of time instructing pupils about them: 'This is how X works; this is how Y moves; and this is the proper way to take Z apart'.

Again we have a dramatically contrasted approach. One cannot help speculating that the middle ground between these two positions is the most fertile for children's learning. Clare and Rogers (1994) for example, have shown the power of the 'process diary' as a tool to help children *to reflect* upon the experiences they have had. This reflection, upon skills and understandings that are being used tacitly, brings them out into the daylight

and helps to make them explicit. It thereby makes them more robust and more transferable. But it does it from the starting point of the child's direct experience. Millar *et al.* (1994) make a similar point about the interaction of children's tacit and explicit understandings in the context of learning science.

The evidence suggests that we should not be satisfied with a purely tacit level of operation, but neither should we ignore children's tacit understandings about how the world works – seeking instead to implant our own 'correct/proper' answers. We should rather work from children's tacit understandings and require them constantly to reflect upon them in the light of the new experiences that we provide for them. Hopefully,

tacit + guided reflection = explicit.

Discussing and progressing

We have outlined in earlier chapters the significance of pupil discussion in helping to develop ideas and organise appropriate lines of action to make progress through a task. Given all that has been said to this point, the reader will hardly be surprised to see that the boundary between Year 6 and Year 7 also provides a very stark contrast between the amount of time spent *listening to the teacher* as opposed to *discussing* work amongst themselves (Fig. 7.6)

It is interesting here to compare these new findings with those from the APU survey. Early trials of the APU tests showed that 15-year olds found it difficult to make effective use of open discussion when they were designing. However, when the discussion was focused around three specific questions it was seen to be much easier and a far more valuable development tool. The questions were:

- explain what you have done up to this point;
- what are the strength and weaknesses of your proposed product;
- what might you do to overcome perceived weaknesses.

The APU discussions were based on two minute presentations to address these three questions followed by a forum/discussion with six pupils

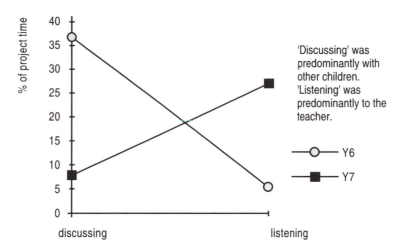

Fig. 7.6 The proportion of time spent discussing and listening

round the table. Following the test, the pupils were asked to comment on the helpfulness of the various strategies used (including the discussion), and it was the discussion that most found to be the most helpful. Table 5.1 (p. 84) illustrates the level at which pupils found the discussion to be helpful.

Our new data shows how little this strategy is currently used in secondary Design and Technology, which also explains why pupils found it difficult to do without a clear structure. We were essentially asking them to operate a strategy that they would probably have had little experience of since they left primary school. As we have pointed out earlier, our fieldworkers who ran the APU modelling tests which included the discussion, spoke with missionary zeal about a strategy that they thought to be very helpful to pupils and (at the same time) virtually unknown in their experience of secondary Design and Technology. Yet it is a standard tool in Year 6 Design and Technology.

An appeal for a big picture

The evidence we have collected of the Design and Technology experiences of children at the top end of primary schools and the bottom end of secondary schools shows clearly the discontinuity of typical primary and secondary activities in technology.

While many of the activities we observed were positive learning experiences for the children, these glaring discontinuities raise serious questions for the further development of Design and Technology.

The problem is (as we pointed out in Chapter 3) that Design and Technology is a young subject in secondary schools and totally new in primary schools. There is too little accumulated experience that allows us to project a coherent model of Design and Technology across the 5–16 age range. We can report that the evidence is that there are huge discontinuities of practice, but the question remains, 'What should be done about this?'

The only option is to take a look at the big picture and then to see how the bits of that picture might best contribute to the whole. This means cross-phase dialogue and planning. Infant, junior and secondary teachers need opportunities to talk to each other to agree priorities and practices. It should be noted that publishers do not help this process by insisting that resource books for teachers are targeted *either* at primary *or* at secondary markets. It may be good business, but it is very bad for the development of any collective understanding about the nature and development of capability.

Unless this issue is tackled, we will continue to perpetuate a very unsatisfactory state of affairs.

In the space of one short summer holiday,

children have completely to reconstruct their mental models about what Design and Technology is like. This may also be true in Maths and English and Science, since many of the differences relate to general teaching/learning differences, but it is glaringly and dangerously true in Design and Technology.

SUMMARY

- We have been struck by the startling differences between technology project work in Key Stage 2 and that in Key Stage 3, and particularly between Years 6–7. We use four of the UTA project case studies in this chapter to highlight some of these contrasts:

    design a museum exhibit for the 21st century
    designing moving vehicles
    designing a plant-watering sensor
    designing an action sports trophy

- All the projects were successful in the eyes of the teachers and in all cases the pupils enjoyed what they were doing.
- Key Stage 2 project work is typified by *uncertainty* and the need to work things out with the teacher.
- Key Stage 3 project work is typified by *certainty* and predictability, under the tight control of the teacher.

- There are major differences in the extent to which teachers *direct* pupils into doing what the teacher sees as important or *supports* them in doing something that they (the children) have decided to do.
- This results in the average working time being reduced for Year 7 pupils, since so much time is taken up in instruction.
- The focus of this instruction is almost entirely on *making* skills.
- Children at Key Stage 2 see the 'user' as important in deciding how their final product should look and operate. At Key Stage 3 the user is not seen as important.
- There is a major change in styles of working, illustrated clearly in the balance of time spent in *discussing* ideas (usually with peers) and in *listening* to the teacher. The patterns for Year 6 are almost exactly reversed in Year 7.
- Activities at Key Stage 2 make great use of the tacit understanding of children (about materials, etc.). At Key Stage 3, teachers are concerned with making these explicit.
- There is a clear need to establish a big picture of capability that spans the four key stages. Only then can we see how each key stage can contribute to the progressive development of capability in Design and Technology.

CHAPTER 8

Helping teachers with the UK National Curriculum

Introduction

Throughout the preceding chapters we have raised and debated a wide range of issues that bear upon the *understanding* of practicc in Design and Technology. We have not thus far attempted to reference these issues into the specific requirements of the UK National Curriculum. This is partly because the arguments raise bigger issues than can be dealt with in the specifics of any National Curriculum Order, and partly because their breadth has enabled us to discuss matters that are (hopefully) informative for colleagues from other nations seeking to develop their own technology curricula.

However, the introduction of the revised National Curriculum in September 1995 (for Key Stages 1–3 and in 1996 for Key Stage 4) provides a challenge for teachers that might usefully be informed by our research findings, and it is to this task that we turn our attention in this final chapter. We have chosen to organise the chapter into four broad sections, covering respectively *generic issues of Design and Technology*, *long-term planning issues*, *medium- and short-term planning issues* and *quality assurance issues*.

- *Generic issues*: affecting the overall understanding of Design and Technology capability (which needs to be shared at the whole school level).
- *Long-term planning issues*: concerned with developing phase and key stage schemes of work (to be debated within departments or with coordinators).
- *Medium/short-term planning issues*: concerned with planning activities and lessons (to be considered by individual teachers).
- *Quality-assurance issues*: concerned with maintaining and developing standards (to be considered by the coordinator/head of department).

Generic issues

As we pointed out in Chapter 3, Design and Technology has grown from practice rather than from a theoretical analysis of a body of knowledge. It developed in classrooms in which teachers tried things out and gradually refined a set of ideas and approaches that contributed something quite new to children's education. It is now firmly established as a subject area in secondary schools and a curriculum focus in primary schools. There has been sufficient success with pupils over an extended period of time for many people to have seen for themselves just how motivating and enriching Design and Technology activities can be for youngsters.

Seeing it working – and even doing it oneself – is not the same as being able to articulate an underlying rationale. It is essential for the further development of Design and Technology that we should be absolutely clear about what good activities look like and what qualities we can expect children to develop as a result of such activities. We focused our attention on these matters in Chapters 2 and 3. Our view, expressed in these early chapters, is that Design and Technology is a purposeful, goal-directed activity that is summed up through simple statements like 'I can see a better way of doing that!'. It is about achieving desires, addressing dissatisfactions, realising visions, both for ourselves and for others. It is also about achieving these things through intervening in the 'made world' of products and systems, using materials and tools to create a better made world. The essence of this view is echoed in our National Curriculum documentation.

Design and Technology in schools

Intervening creatively to improve the made world

In Chapter 3, we considered the reasons for having Design and Technology as a curriculum activity. It is important to identify its unique contribution, what it is that separates it from Science or History or Music. We identified one of these distinctive features residing in the fact that Design and Technology is concerned with more than just understanding the world; it is concerned with making use of knowledge and skill to enable us to *intervene creatively* so as to improve the made world in response to our needs.

This feature of the activity was far better described in the 1990 Order for Design and Technology than it is in the current 1995 Order, which restricts itself to the requirement of engaging in 'assignments in which pupils design and make products' and further requires (through the programmes of study) that pupils should be taught, for example to 'generate ideas, considering the users and purposes for which they are designing'. However, the 1995 Order is rather more helpful in exemplifying the three kinds of activity that we describe as contributing to *awareness* of technology, *competence* in technology and *capability* in technology.

For each key stage, these three distinct types of activities are required:

- *Activities in which pupils investigate, disassemble and evaluate products and applications.* Those

which support pupils' awareness of what Design and Technology is and how it can impact on people's lives. In addition, these types of activities can increase competence in Design and Technology by providing opportunities to develop knowledge and understanding of materials, manufacturing processes, finishing processes, etc.
- *Focused practical tasks in which they develop and practise particular skills and knowledge.* Those which will help pupils become competent in making things work.
- *Assignments in which pupils design and make products.* Those which will allow them to develop capability, drawing together knowledge, skill and understanding and utilising these in creating new products that meet specific needs.

We describe this final level as 'not so much studying technology as being a technologist' and we identify the aim of developing Design and Technology *capability*, which is 'dependent upon a combination of abilities and motivations that empower us to bridge the gap between human aspiration and technical constraint'. The professional associations, such as National Association of Advisers and Inspectors of Design and Technology (NAAIDT) and Design and Technology Association (DATA) have also identified the value of this triumvirate:

> DATA sees these three types of activity as being inter-related, with IDEAs (Investigative, Disassembly and Evaluative Activities) and FPTs (Focussed Practical Tasks) supporting and enriching DMAs (Design and Make Assignments). In a sense the IDEAs and FPTs involve and ensure focussed teaching within a DMA, which helps to develop progression in pupils' learning.
>
> (DATA, 1995)

However, we also identify in Chapter 4 the essential similarities of these three kinds of activity. We suggest that 'focused tasks merge imperceptibly with DMAs as the focus gets progressively less sharp'. This is an important issue, since we show in Chapter 4 that the differences between these kinds of task in Design and Technology are explained by the stance that teachers take in respect of three characteristics:

- the balance between the product purpose and the teaching purpose of the tasks;
- the relationship between particularised tasks and generalised contexts; and
- the balance between teacher control and pupil autonomy.

Any task will be definable in terms of teachers' position in relation to these three and, by modifying that position, it becomes possible to transform the nature of the task. What this suggests is that teachers do not have only three kinds of task available to them, but rather endless shades of difference.

We also identify, with the evolution of Design and Technology, a shift in emphasis from outcome to process. This is not to deny the importance of the outcome but to amplify 'a move from receiving "hand-me-down" outcomes and truths to one in which we generate our own truths. The pupil is transformed from passive recipient into active participant'. Further, we distinguish two types of outcome from a Design and Technology 'capability' course: the *product* outcome which the pupil has designed and made, and the *learning* outcome in terms of the pupil's enhanced capability in designing and making.

Both the 1990 and 1995 Orders assess the attainment of capability in procedural terms:

1990	1995
AT1: Identifying needs and opportunities	
AT2: Generating a design	AT1: Designing
AT3: Planning and making	AT2: Making
AT4: Evaluating	

All the level descriptions for attainment within the 1995 Order begin with the words 'when designing and making . . .'. Process-centred capability, the ability effectively to pursue an activity from inception to completion, is the central requirement of Design and Technology capability. It requires the development of knowledge and skill, but that is not the point of it all. The point is to be able creatively

to make use of that knowledge and skill in tackling tasks in the made world. This is the first central requirement that lies at the heart of the 1995 Order.

Values and Design and Technology

A second key aspect of Design and Technology is thrown up by the notion of 'improvement' in our view of Design and Technology. To say 'I can see a better way of doing that!' raises the question, 'Better for whom?' and from this comes the realisation that rarely is any innovation better for everyone. Design and Technology is heavily embroiled with our values and priorities, and decisions require us to confront a whole host of (often conflicting) issues: moral, technical, aesthetic and economic. Prioritising and optimising therefore become essential qualities. This dimension of Design and Technology further enriches it as a learning area, as involvement in handling these issues has the double value of raising children's awareness of the issues themselves while also (since Design and Technology demands action) helping them to develop strategies for dealing with the issues.

In Chapters 4 and 5 we accentuate the importance of the values dimension of Design and Technology, both in terms of encouraging pupils to take ownership of, and responsibility for, their tasks and in helping them to pursue their work through addressing both 'user' and 'making' issues in their designs. This position is clearly outlined in the programmes of study in the 1995 Order, where, from Key Stage 1 onwards, it is required that children be taught to view a product from the point of its purpose, how it meets people's needs and what users think of the product. This starting point is then built on so that by Key Stage 3 the programme specifies, for example, that pupils be taught that the quality of a product relates at the same time to its fitness for purpose, its appropriateness in terms of use of resources and its 'impact beyond the purpose for which it was designed, e.g. the environment.' In other words, Design and Technology activities become much more meaningful for children if they see them as involving the design and manufacture of *something*, for *some purpose*, for *someone*.

Considering values is a critical part of Design and Technology and it follows that teaching and learning about values, and experiencing the difficulty of optimising different value positions, is a critical dimension of Design and Technology in the curriculum.

Imaging and modelling – creative concrete thinking

A third aspect of Design and Technology which we have returned to again and again through both the text and the examples of children's work, is the way in which pupils engage with the activity through 'hands-on', creative, concrete thinking. Pupils design – and think – through materials, they handle value issues – through materials, they develop conceptual understanding – through materials. Central to this are two related dimensions: their ability *to image and model ideas* and their ability *to reflect and take action* to turn those ideas into working realities. We have seen through examples of children's engagement with Design and Technology just how powerful this approach to learning can be. But equally we have seen how debilitated a pupil can be from being denied the opportunity to work in this way.

The balanced development of active and reflective dimensions of capability is particularly important, as we pointed out in Chapter 3 (see pp. 29–32).

The very fact that National Curriculum Design and Technology is so strongly centred on the process of designing and making makes it easy to adopt the model of action and reflection as the principal mode for teaching and learning. The programmes of study inherently specify such an approach through the combination of demands they contain:

'generate ideas considering the users . . .'
(PoS Designing Key Stage 2)
'evaluate their products as these are developed. . .' (PoS Designing Key Stage 1)
'implement improvements they have identified.'
(PoS Making Key Stage 2)

'consider the physical and chemical properties of materials and . . . *relate these* . . . to the ways materials are worked and used.' (PoS Materials and Components Key Stage 3)
'modify their proposals in the light of on-going analysis and product development.' (PoS Designing Key Stage 4)

(DFE, 1995) [italics added]

In each case, the action is simultaneously linked to reflection upon it and similarly, through the descriptions of attainment, the same pattern of linked action and reflection is repeated:

'. . . pupils generate ideas, *recognising that their designs will have to satisfy conflicting requirements* . . .' (AT Level 3 Designing)
'. . . pupils produce plans that outline *the implications of their design decisions* . . .' (AT Level 6 Making)

(DFE, 1995) [italics added]

Learning to take action through the concrete world of materials, whilst at the same time being able to stand back and reflect on that action, is the third central component of capability.

It is these three features of the activity which are important for all teachers of Design and Technology. They are the central defining qualities that set Design and Technology apart from the rest of the curriculum and, in order that all involved staff are aware of the overarching nature of these issues, we believe that they should therefore be addressed when subject policies and guidelines are being written.

Long-term planning issues

There are a number of concerns that we have raised in the preceding chapters that bear directly on what we would class as long-term planning. These are broad concerns, for example of progression and how it is accommodated into schemes of work for whole phases or for key stages. This is large-scale, broad-brush planning that can subsequently lead to the mapping of an appropriate set of activities.

Progression and transition

We have identified in Chapter 5, a range of qualities that we believe to be central to the development of capability. Moreover, we have exemplified what these qualities look like at each of the four key stages using the same observation framework. Accordingly, we can glimpse (perhaps for the first time) what it might mean to talk about progression in these aspects of pupils' learning.

We have included in our list qualities that are procedural, those to do with knowledge and understanding, and those that centre on making skills and communicating skills. If children's Design and Technology capability is to progress, there has to be progression in each of these individual skills and abilities, and the 1995 Order provides very little guidance for teachers trying to grapple with this issue. We believe that teachers need to develop mental 'pen-pictures' of what, for example 'investigation' should consist of at Year 6. These pictures will emerge from a combination of the statutory requirements (e.g. outlined in the Order and in other official documents), combined with exemplification materials of the kind we have used in Chapter 5, and using teachers' personal experience of running projects. It will then be possible to see how that Year 6 'standard' represents an advance on what Years 2 and 4 investigating might consist of. From this will gradually emerge a more comprehensive picture of progression as a whole.

In Chapter 7 we have highlighted some of the gross discontinuities that exist across the key stages. Particularly at the Year 6–7 boundary, an outside observer might assume that teachers were taking an almost perverse delight in contradicting every single element of each other's teaching approaches. This kind of discontinuity is damaging for children and can only be eliminated if we take a big picture overview of what each key stage can contribute to the whole development of capability.

We must, therefore, debate the purpose of tasks, the nature and level of task setting and the degree of autonomy that pupils will have in pursuing them. We characterised this as a contrast between certainty and uncertainty at the Year 6–7 bound-

ary. It also requires us to consider how exposure to the rich diversity of materials and manufacturing techniques that is typical of Key Stage 3, can be made less instructional. It needs to feed more readily from the teaching styles that predominate at Key Stage 2. There has already been some suggestion that Key Stage 2 teaching might be modified somewhat to become more specialist and more like Key Stage 3. We would suggest that this should be a double-ended modification, with changes to both such that the interface at Year 6-7 is less traumatic. If this is to occur there needs to be consideration of how the subject knowledge and expertise of the primary teacher can be enhanced, and how the secondary teachers' management of learning can draw on the best primary practice.

Consideration of these points on initial teacher training and in-service courses, along with improved links between secondary schools and their partner primaries, would all help to establish a more coherent and progressive development of capability. We suggest this not just out of concern for the feelings of children, but rather from the conviction that there has to be a better way of progressively developing capability than doing it one way for four years in Key Stage 2 – and the reverse way for the next three in Key Stage 3.

We have drawn attention, particularly in Chapter 7, to the contrasted qualities that these different teaching regimes develop in youngsters. We would argue that the change from one key stage to the next might sensibly result in a change of emphasis in children's work and how it is organised. But the changes should not be so dramatic that they undo all the good work of previous years. They should rather build upon it.

In reconciling these broad-brush issues of keystage planning, we would expect schools to have a view about the range of materials that might be used and the range of production technologies that might reasonably be introduced. Equally, we would expect these plans to demonstrate the diversity of contexts within which tasks are set, in order that during a key stage, pupils design and make in a *range* of situations and avoid too narrow a focus in the work.

Medium- and short-term planning issues

Once the overall framework for the key stage has been established – hopefully in discussion with those responsible for adjacent key stages – it is necessary to move to a finer level of detail. Here we can begin to focus on the specifics of what kinds of activities might be used and what demands they should properly make on children's developing capability.

Devising tasks

In Chapters 4, 5 and 6 we outlined a series of issues that bear directly on the nature of tasks and on the consequences of using them in one form as opposed to another. These 'task effects' result in a series of recommendations that might be summarised as follows. In devising tasks we should seek to ensure the following are taken into account:

- Tasks relate to an identifiable *client*. This helps them to focus the 'user' issues as well as the technical ones and encourages the evaluation of outcomes in terms of fitness for purpose.
- Tasks are set that have a range of *value positions* within them (technical, aesthetic, moral/ethical, economic). This will encourage pupils in their optimising of issues and priorities and help them to see how this results (in outcome terms) in winners and losers.
- The *entry point* into tasks should recognise that there is a hierarchy that stretches from the generalised context to the particularised task. Tight tasks can be supportive, but difficult to personalise to a real user. Loose tasks can be easy to personalise, but can allow children to lose any focus on their learning. We should encourage a balance of such entry points so that pupils become familiar with the different challenges of handling tasks at any level.
- The starting point for tasks places appropriate levels of responsibility upon pupils. Teachers need to be aware of how their activities can develop or suppress the child's autonomy.
- The starting point for tasks will affect to a

marked degree the extent to which pupils can take 'ownership' of tasks. It is clearly desirable that they should see the task as important not just as 'a school exercise' but also as a real task that has real consequences for the made world, bearing in mind their age and experience. If teachers can help children to take ownership of the tasks, they can also expect more of the pupils (since the pupils will be driven not just by the teacher) but also by the imperative of the product outcome.

- We should be aware that all these variables will be perceived differently by different sub-groups of children. All children are different and require different elements of support or challenge. We explored this particularly in Chapter 6.

Structuring activities and supporting pupils

Equally, however, we have considered in Chapters 4, 5 and 6 how the structure of sub-tasks through the activity can affect not only the ease with which children can tackle it, but also the extent of the challenge within it.

- The language of Design and Technology is a concrete one, and tasks should be structured to provide opportunities for concrete modelling in different ways; the use of drawings, paper modelling, clay/'Plasticine' modelling, verbal modelling (discussion), 3D mock-up modelling.
- These modelling opportunities each allow children to explore and express their ideas in different ways. They are therefore essential to the extension of children's creative concrete thinking.
- The *action* of modelling encourages children's *reflection* on the consequences for their developing solution. It is one of the cornerstones of Design and Technology that we can help pupils to make their thought processes explicit, exposing these processes to the light of day. Their ideas folders (of drawings) will help this process, as will the models, the progress reviews and the formal evaluations.
- The importance of *discussion* cannot be overstressed, since it makes such a contribution to

the pupil's development of capability. We have shown in Chapter 5 just how vital it can be.

- All teachers are aware of the strengths and weaknesses that individual children possess. But part of developing capability requires us to strengthen our weaknesses rather than always playing safe with our strengths. Activities need to encourage children to work *from* their strengths, not *on* their strengths.

Quality assurance issues

We have described in Chapter 3 some of the issues of quality that have arisen in recent years in the context of teaching Design and Technology. Too often it has been assumed that 'quality' is something that can be written down in a statutory document and thereafter applied like a ruler. This is a serious misconception, since quality cannot ultimately be determined by agencies or through publications that are external to the school. Just as a 'quality' restaurant is not defined by the fact that it uses an *haute cuisine* cook-book. Quality resides in the professionalism and the gifts of the participants; the chef and the teacher.

We can of course help to inform this professionalism: through initial training; through continuing professional development; through distance learning support from curriculum development agencies; and (sometimes) through statutory documents like the National Curriculum Orders. But these Orders should not be seen as defining practice since they say almost nothing about *how to do things* in the classroom. They concentrate instead on the outcomes that should be sought.

Models of 'quality control', for example including practices like 'inspections' tend to focus on these outcomes. In industrial terms, it is a bit like waiting until the end of the production line before rectifying the fault or throwing away the faulty component. Industrial practice has long since moved away from this ineffective and costly approach and adopted instead a 'quality assurance' model where quality is constantly checked and faults rectified as they occur in the workplace. Quality *control* assumes that idiots or robots are responsible for production. Quality *assurance* assumes that professional practitioners are responsible.

Quality assurance, therefore, requires professionalism and involves setting agreed targets and determining ways of working towards them. It is the fact that quality assurance is integral to the professional practice of teaching that makes it a more appropriate model for education, and this is increasingly being pursued through a variety of channels.

Even though there is now meant to be a temporary halt to curriculum change, there is unlikely to be a slow down in the rate at which schools are offered advice on how they might best move forward. In secondary schools this has largely revolved around issues such as raising attainment at 16+, league tables, value-added performance indicators, coursework percentages, 'A'-levels, GNVQs, etc. In the primary sector, the debate has centred on methodology, content and subject management. In terms of promoting quality teaching and learning, it is the role of the subject co-ordinator that has become the focus of attention:

> In practice, while co-ordinators have often had an impact upon both whole school curriculum planning and the management of resources, in many schools they have had little real influence on the competence of individual teachers and the quality of classroom teaching and learning.'
> (Alexander *et al.*, 1992)

The follow-up report to this discussion paper also made similar comments (OFSTED, 1993c). But it was the publication of *Primary Matters – A Discussion on Teaching and Learning in Primary Schools* that made recommendations as to how the quality assurance role of the co-ordinators might change. And it might equally be seen to apply to the roles of heads of department/faculty at secondary level.

> In the majority of schools curriculum co-ordinators engaged in several of the following activities: writing policies and schemes of work; leading meetings of staff; providing INSET; attending courses; auditing, purchasing and organising resources; advising teachers; and supporting or leading planning. Very few had a role which

extended to monitoring or evaluating the quality of the work in their subjects. In a few schools the co-ordinators were released from their own classes to work alongside colleagues, but this was usually planned as a support rather than a monitoring activity . . .

Headteachers are able to delegate the management of particular subjects to individual members of staff . . . teachers who are subject managers . . . 'co-ordinators' is too limited a description . . .

Subject managers can also be expected to contribute to the overall evaluation of work in their subject against agreed criteria, to evaluate standards of achievements; and to identify trends and patterns in pupils' performance. Such contributions to evaluation are more valid if informed by the first hand experience of monitoring work in classes but can also be based, at least in part, on the examination of the teachers' and children's work.

(OFSTED, 1994b)

This could be summarised by saying that the co-ordinator in primary schools, and the head of department/faculty in secondary schools, needs continually to ask four questions.

- Do we all share common standards of quality?
- Is the planned curriculum delivered?
- Have standards improved in the subject area?
- How might we modify what we are doing to improve our effectiveness?

Overriding the answers to these questions is perhaps an even more important one, 'How do I know?'

The development of common standards of quality in Design and Technology has to be addressed at the whole-school level in primary schools and at department/faculty level in secondary schools. Sharing and agreeing what capability consists in has to be the basis for any development in the subject. Chapter 5 detailed our findings and interpretations of the data collected in the UTA study and the findings should be helpful to teachers in formulating their own views of standards.

Once these yardsticks of quality have been established, it is possible to plan how they might be implemented, and the long-, medium- and short-term planning issues highlighted earlier in this chapter come into play. Monitoring the 'delivery' of the resulting curriculum then becomes a key consideration in assuring quality. The coordinator/head of department needs constantly to be gathering data about the performance of teachers and pupils in the subject, and from a variety of sources. She or he needs to understand why X is doing Y, and what the consequences are for Z. The evidence needs to go further than hunch and instinct. It needs to address the programmes of study, the activity schedules and the observation of children at work on tasks. The collection of this information over a period of time allows teachers to address and answer those critical questions of quality, and whether or not improvement is occurring. The information has to be seen as providing the means through which professional debate can take place and is aimed at ensuring an improved Design and Technology experience for all pupils. If this is seen to be intrusive by some colleagues, we should remember who we would want to be the guardians of quality in our schools. Do we want to be responding (like idiots and robots) to externally imposed inspection, or would we rather be developing models of quality from within, capitalising on the professionalism of our teachers?

Teaching Design and Technology is a difficult task. It is relatively new in the curriculum and its statutory introduction in the National Curriculum has created a number of problems for schools. It is these problems that we have sought to illuminate through this book:

- Problems of definition: what is Design and Technology?
- Problems of practice: how do we teach it?
- Problems of progression: how do children learn it?

Through all these difficulties, teachers should remember why they are doing it and why it has been so valuable and so enjoyable for so many children in so many schools.

Conclusion

Design and Technology in the school curriculum has grown from *practice* rather than from theory; from teachers in classrooms trying out innovative and often idiosyncratic activities and programmes rather than from an intellectual analysis of a field of knowledge. And it has been hugely successful. Pupils voted with their feet; courses expanded and proliferated; competitions and prizes led to high profile public exposure where politicians and others were delighted to shake a few hands for the camera. Eventually, with the growth of Advanced level work, even the universities caught up with the fact that there were some quite exceptional young talents coming through this route and increasingly sought them out. This growth has been steered by national projects, fostered by national, regional and local curriculum initiatives, and has gradually been disseminated to a wider teaching force.

But the roots of its success lie in individual classrooms, studios and workshops where imaginative teachers sought to create not just a new subject – a new approach to learning. Technology is now coming of age and has been made a compulsory study for all pupils. But the confusions that have surrounded this move bear ample testimony to the difficulty of making the transition. Design and Technology was developed by enthusiasts and taken forward by converts, many of whom demonstrated a semi-religious conviction about its value. The problem presented by the National Curriculum was that this conviction had to be tamed and institutionalised so that every teacher could cope with it. And the first step towards this required that we write down (in a National Curriculum Order) what it is that we do.

We should not be surprised that this process proved difficult. It was difficult because ultimately technology cannot be defined in a body of knowledge and skills. It is better described as a set of ideals and processes into which pupils need progressively to be drawn. Naturally this induction relies upon the pupil developing knowledge and skill, but ultimately, pupils are not to be judged by their mastery of these skills so much as by their ability to recognise and grasp opportunities to change and improve the world through the exercise of their creative talents. This was the big idea that Bronowski (1973) chronicled in *The Ascent of Man* and that lay at the heart of those pioneering courses in the 1970s and 1980s.

Another part of the difficulty of developing Design and Technology has been the paucity of research to inform our collective understanding about the endeavour. We would like to think that two exceptions to this general rule are the two research projects conducted in the Technology Education Research Unit at Goldsmiths College since the mid-1980s, and on which this book is based; i.e. the APU project (concerned with assessment) and the UTA project (concerned with models of teaching and learning). These projects form the backbone of this book by providing the data – the experience – from which we have constructed our current understandings of classroom practice and pupils' performance.

Inevitably, these projects have raised as many questions as they have answered – but that is the nature of research. Nevertheless, we hope that this book will make a contribution to the further development of Design and Technology specifically through our attempts to use our research data to address the following four questions:

- What is Design and Technology?
- How do we teach it?
- How do children learn it?
- What is its unique role in the curriculum?

And in the process of addressing these questions we have been constantly reminded of a central truism about Design and Technology – which is that children enjoy it.

Bibliography

Allen, Sir G. (1980). Engineering: Finniston and the future, *Times Higher Education Supplement*, 10 October, p. 1.

Alcxander, R., Rose, J. and Woodhead, C. (1992). *Curriculum Organisation and Classroom Practice in Primary Schools – A Discussion Paper*. London, DES.

Archer, B. and Roberts, P. (1992). Design and technological awareness in education. In: *Modelling: The Language of Designing*. Occasional Paper No. 1. Loughborough, Loughborough University of Technology.

Baynes, K. (1986). Designerly play, *National Association for Design Education Journal*, Spring.

Baynes, K. (1989). The basis of designerly thinking. In: *Looking, Making and Learning: Art and Design in the Primary School*. Dyson, A. (ed.). London, Kogan Page.

Baynes, K. (1992a), *Children Designing*. Occasional Paper. Loughborough, Loughborough University of Technology.

Baynes, K. (1992b). The role of modelling in the Industrial Revolution. In: *Modelling: the Language of Designing*. Occasional Paper No. 1. Loughborough, Loughborough University of Technology.

Baynes, K. and Pugh, F. (1981). *The Art of the Engineer*. Cambridge, Lutterworth Press.

British Broadcasting Corporation (1994). The future is female. *Panorama*, 24 October.

Bronowski, J. (1973). *The Ascent of Man*. London, British Broadcasting Corporation.

Bruner, J. (1968). *Towards a Theory of Instruction*. New York, W. W. Norton.

Carraher, T. N., Carraher, D. W. and Schliemann, A. D. (1985). Mathematics in the streets and in the schools, *British Journal of Developmental Psychology*, 3: 21–9.

Clare, D. and Rogers, M. (1994). The process diary: developing capability within National Curriculum design and technology – some initial findings. In *IDATER 94*. Smith, J. S. (ed.). Loughborough, Loughborough University of Technology.

Consumers Association (1992). Gas and dual-fuel cookers, *Which?* August.

The Corfield Report (1976). *Product Design*. London, National Economic Development Office.

Darke, J. (1979). The primary generator and the design process, *Design Studies*, 1(1): 36–44.

Design and Technology Association (DATA) (1995). *Guidance Material for Design and Technology*. London, DATA.

Davies, D. (1994). Professional Design and Primary Children. MA dissertation. Goldsmiths University of London.

Denton, H. G. (1993). Designing efficiently and effectively: do we encourage children to use 'design sheets' appropriately?, *Design and Technology Teaching*, 25(3): 48–52.

Derricott, R. (1985). *Curriculum Continuity: Primary to Secondary*. Windsor, NFER Nelson.

Department for Education (1995). *Design and Technology in the National Curriculum*. London, HMSO.

Department of Education and Science/Welsh Office (1987). *The National Curriculum 5–16: A consultation Document*. London, Department of Education and Science and Welsh Office.

Department of Education and Science/Welsh Office (1988). *National Curriculum Design and Technology Working Group – Interim Report*. London, DES.

Department of Education and Science/Welsh Office (1990). *Design and Technology for Ages 5–16*. London, HMSO.

Department of Education and Science/Welsh Office

(1992). *Technology Key Stages 1, 2 and 3. A Report by HMI on the First Year 1990–91.* London, HMSO.

Department of Education and Science/Welsh Office (1995) *National Curriculum Design and Technology Statutory Order.* London, HMSO.

Desforges, C. (1995). *Introduction to Teaching: Psychological Perspectives.* Oxford, Blackwell.

Dewey, J. (1968). *Democracy and Education.* New York, Free Press.

Donaldson, M. (1978). *Children's Minds.* London, Fontana.

Downey, M. and Kelly, A. V. (1979). *Theory and Practice of Education.* London, Harper & Row.

Eggleston, J. (1992). The politics of technology education. In: *Technological Literacy. Competence and Innovation in Human Resource Development. Proceedings of the first International Conference of Technology Education.* Weimar, Technical Foundation of America.

Finniston, Sir M. (1980). *Engineering our Future.* London, HMSO.

Galton, M. and Wilcocks, J. (1983). *Moving from the Primary Classroom.* London, Routledge & Kegan Paul.

Gorman, M. and Carlson, B. (1990). Interpreting invention as a cognitive process: The case of Alexander Graham Bell, Thomas Edison, and the telephone. In: *Science Technology and Human Values,* 15(2): 131–164.

Hennessy, S. and McCormick, R. (1993). The general problem-solving process in technology education. In: *Teaching Technology.* Banks, F. (ed.). London, Routledge.

Humphreys, G. (1951). *Thinking.* London, Methuen.

Inholder, B. and Piaget, J. (1958). *The Growth of Logical Thinking from Childhood to Adolescence.* New York, Basic Books.

Johnsey, R. (1995). Criteria for success, *Design and Technology Teaching,* 27(2): 37–9.

Kelly, V., Kimbell, R. A., Patterson, V., Saxton, J. and Stables, K. (1987). *Design and Technological Activity. A Framework for Assessment.* London, HMSO.

Kimbell, R.A. (1982). *Design Education: The Foundation Years.* London, Routledge & Kegan Paul.

Kimbell, R. A. (1991). Tackling technological tasks. In: *Practical Science,* Woolnough, B. (ed.). Buckingham, Open University Press.

Kimbell, R. A. (1992). Technology: a problem of success, *Times Education Supplement,* 16 October, p. 6.

Kimbell, R. A. (1994a), Tasks in technology: an analysis of their purposes and effects, *International Journal of Technology and Design Education,* 241–56.

Kimbell, R. A. (1994b). *Understanding technological approaches: Final report of research activities and results.* R-000-23-3643. London, Goldsmiths College, ESRC.

Kimbell, R. A. (1994c). Gender differences in technology. In: *Design Studies.* London, Butterworth-Heinemann.

Kimbell, R., Stables, K., Wheeler, T., Wozniak, A. and Kelly, V. (1991). *The Assessment of Performance in Design and Technology: The final report of the Design and Technology APU project.* London, Evaluation and Monitoring Unit. Schools Examination and Assessment Council (SEAC).

Kimbell, R. and Wheeler, T. (1991a). *Negotiating Tasks in Design and Technology.* London, SEAC/HMSO.

Kimbell, R. and Wheeler, T. (1991b). *Structuring Activities in Design and Technology.* London, SEAC/HMSO.

Kubie, L. (1962). Blocks to creativity. In: *Explorations in Creativity.* Mooney, R. and Razik, T. (eds). London, Harper & Row.

Langer, S. (1962) *Philosophical Sketches.* Johns Hopkins University Press. Baltimore, MD.

Layton, D. (1993). *Technology's Challenge to Science Education.* Buckingham, Open University Press.

Lynch, I. (1993). *Technology Science Education and the World of Work.* London, City Technology Colleges Trust.

McCormick, R. and Murphy, P. (1994). Learning the process of technology. Paper presented to the British Education Research Association annual conference, Oxford, September 1994.

Millar, R., Lubben, F., Gott, R. and Duggan, S. (1994). Investigating in the school science laboratory: conceptual and procedural knowledge and their influence on performance, *Research Papers in Education,* 9(2): 36–52.

Munn, P. (1995). What do children know about reading before they go to school? In: *Children Learning to read: International Concerns: Vol 1.* Owen, P. and Pumpfrey, P. (eds). London, Falmer Press.

National Association of Advisers and Inspectors in Design and Technology (1993). *Quality in Design and Technology: what we should be looking for.* NAAIDT. Also published in *Design and Technology Teaching,* 26(2): 53–5.

National Association for Science Technology and Society (1991). *Sixth Annual Conference of the National Association for Science Technology and Society.* Washington DC, NATS.

National Curriculum Council (1993). *Quality in Design and Technology.* London, NCC.

Noble, D. (1984). *Forces of Production: A Social History of Industrial Automation.* New York, Alfred A. Knopf.

Office for Standards in Education (OFSTED) (1993a). *Technology: Key Stages 1, 2 and 3: A report from the Office of Her Majesty's Chief Inspector of Schools, 1991–92.* London, HMSO.

Office for Standards in Education (OFSTED) (1993b). *Technology: Key Stages 1, 2 and 3: A report from the Office of Her Majesty's Chief Inspector of Schools, 1992–93.* London, HMSO.

Office for Standards in Education (OFSTED) (1993c). *Curriculum organisation and classroom practice in primary schools – a follow-up report.* London, HMSO.

Office for Standards in Education (OFSTED) (1994a). *Handbook for the Inspection of Schools.* London, HMSO.

Office for Standards in Education (OFSTED) (1994b). *Primary Matters – A Discussion of Teaching and Learning in Primary Schools.* London, OFSTED.

Pacey, A. (1983). *The Culture of Technology. Oxford,* Blackwell.

Penfold, J. (1988). *Craft Design and Technology: Past Present and Future.* Stoke-on-Trent, Trentham Books.

Polanyi, M. (1958), *Personal Knowledge.* London, Routledge and Kegan Paul.

Rogoff, B. (1990) *Apprenticeship in Thinking: Cognitive Development in a Social Context. Buckingham,* Oxford University Press.

Ross, C. and Browne, N. (1993). *Girls as Constructors in the Early Years.* Stoke-on-Trent, Trentham Books.

Schools Council Design and Craft Education Project (1972). *Certificate of Secondary Education: Studies in Design.* Keele University, Schools Council Publications Co.

Science Policy Research Unit (1984). *Project SAPHO – a report.* Brighton, University of Sussex.

Secondary Examinations Council (SEC) (1985). Craft Design and Technology GCSE; A guide for teachers. Buckingham, Open University Press.

Skolimowski, H. (1966). The structure of thinking in technology, *Technology and Culture,* 7: 371–83.

Smithers, A. and Robinson, P. (1992). *National Curriculum Technology – Getting it Right.* London, Engineering Council.

Stables, K. (1992). The role of fantasy in contextualising and resourcing design and technological activity. In *IDATER '92,* Smith, J. S. (ed.) Loughborough, Loughborough University of Technology.

Stables, K. (1995). Discontinuity in transition; pupils experience of technology in Year 6 and Year 7, *International Journal of Technology and Design Education,* 5(2): 152–69.

Stables, K. *et al.* (1991). *CATS Technology: Phase One Evaluation Report.* London, Schools Examination and Assessment Council, Goldsmiths College.

Thomson, R. (1959). *The Psychology of Thinking.* Harmondsworth, Pelican.

Vygotsky, L. S. (1962). *Thought and Language.* Haufman, E. and Vakar, G. (eds/trans.). Cambridge, MA, MIT Press.

Vygotsky, L. S. (1966). Psychological research in the USSR. In: Light, P., Sheldon, S. and Woodhead, M. *Learning to Think* (1991). Buckingham, Open University Press.

Williams, R. (1965). *The Long Revolution.* Harmondsworth, Penguin.

Index

TEACHING DESIGN AND TECHNOLOGY
(SECOND EDITION)

John Eggleston

Reviews from first edition:

> . . . this book deserves a wide readership. Practising teachers should find plenty to interest them. It looks like a clear favourite for the reading lists of students on initial teacher training courses. . . .
>
> *Design and Technology Teaching*

> This book is very readable, besides being an important source of reference. It should be part of any induction course for design and technology teacher training.
>
> *Design and Technology Times*

Design and technology is one of the fastest growth areas of the contemporary school curriculum. It is crucial to the national economy and to individual employment prospects. John Eggleston shows how this area of work has come to occupy a new and central place in the school curriculum, and in doing so has acquired a higher status and a new identity. He explores this new identity, its origins, its manifestations in classroom practice, and its possible futures. He pays particular attention to its role in the national curriculum, to assessment, to gender and race issues, and to management; and concludes with a number of invaluable case studies of school practice.

This best-selling book has been fully revised in the light of the new orders for National Curriculum Design and Technology and is still the only book in the field to examine the subject at a professional level rather than simply providing pupil activities. . . .

Contents

The coming of design and technology – What is design and technology? – Design and technology in the National Curriculum – Assessing design and technology – Gender, race and design and technology – Managing design and technology – Design and technology in practice – Conclusion – Index.

c.128pp   0 335 19502 4 (Paperback)

TECHNOLOGY'S CHALLENGE TO SCIENCE EDUCATION

David Layton

The book explores the relationship between science and technology in the school curriculum. In the past, science has used technological applications to make scientific concepts and ideas more understandable (science and applications). It has also taken technological applications and then extracted the science involved in them in order to make the learning of science more interesting and effective (science of applications). In both cases technology is serving the needs of science education.

With the incorporation of design and technology as a component of the general education of all children, there arises a new situation. Science has now to serve the needs of technology and act as a resource for the development of technological capability in children. However, science for applications is not the same as science and applications or science of applications. Often the science of traditional lessons needs to be reworked to make it useful in practical situations and related to design parameters. Examples of science as a resource for technological capability are drawn from both 'real world technology' and from 'school technology' to illustrate what needs to be done if technology's challenge to science education is to be met.

Contents
The emergence of technology as a component of general education – Technology in the National Curriculum of England and Wales – Understanding technology: the seamless web – Understanding technology: values, gender and reality – Science as a resource for technological capability – Reworking the school science-technology relationship – Responses and resources: a review of the field – Index.

80pp   0 335 09958 0 (paperback)   0 335 09959 9 (hardback)